DOES EVOLUTION EXPLAIN EVERYTHING ABOUT LIFE?

Answers from Ph.D. Scientists

Philip Bell (Contributing Editor)

Ph.D. Scientists:
Jonathan Sarfati | Pierre Jerlström | Robert Carter
David Catchpoole | Stuart Burgess | Tasman Walker
Jim Mason | Donald Batten | Peter Borger | Samuel Gan

Response to the *New Scientist* book,
"How Evolution Explains Everything About Life"

DOES EVOLUTION EXPLAIN EVERYTHING ABOUT LIFE?

Answers From Ph.D. Scientists

Philip Bell (Contributing Editor)

Jonathan Sarfati | Pierre Jerlström | Robert Carter
David Catchpoole | Stuart Burgess | Tasman Walker
Jim Mason | Donald Batten | Peter Borger | Samuel Gan

First printing: 2020

ISBN: 978-1-942773-79-5

Editorial assistance: Donald Batten
Proof reading: Lucien Tuinstra and Gavin Cox
Cover design and layout by: Sarah Fretwell
Published by:

CREATION
BOOK PUBLISHERS

Creation Book Publishers, P.O. Box 350, Powder Springs, GA, 30127, USA.
Phone: 1-800-616-1264 · creationbookpublishers.com

CREATION.com

Please visit our website for further information on the Christian worldview and the creation/evolution issue.

Contents

About the Authors

Philip Bell is CEO of *Creation Ministries International (UK/Europe)*. Formerly a cancer researcher, then a science teacher (11–18 year olds), he is author of *Evolution and the Christian Faith: Theistic Evolution in the Light of Scripture.*

Ph.D. Contributors

Jonathan Sarfati has a Ph.D. in spectroscopy (physical chemistry) and works for CMI-US as a senior scientist and speaker. He is also a chess master. A best-selling author, examples of his books are *Refuting Evolution, By Design* and *The Greatest Hoax on Earth?* (refuting Richard Dawkins).

Pierre Jerlström has a Ph.D. in molecular biology and works for CMI-Australia as a scientist and editor. He is the Editorial Coordinator of the influential *Journal of Creation.*

Robert Carter has a Ph.D. in marine biology and works for CMI-US as a senior scientist and speaker. His contributions to genetics research are published in both creationist and secular journals. He is the editor and a contributor of the acclaimed book *Evolution's Achilles' Heels.*

David Catchpoole has a Ph.D. in plant physiology and works for CMI-Australia. A prolific writer of articles on a diversity of subjects, his co-authored *The Creation Answers Book* is a best seller. He also contributed to *Evolution's Achilles' Heels.*

Stuart Burgess is Professor of Mechanical Engineering at Bristol University (UK). A biomimetics expert, he has won prestigious awards in science and technology. His books include *Hallmarks of Design* and *Inspiration from Creation*.

Tasman Walker has a doctorate in mechanical engineering and used to work in the power station industry. He works for CMI-Australia as a writer and speaker and holds a first class degree in earth science, on which he has written extensively.

Jim Mason has a Ph.D. in experimental nuclear physics. Retired after 37 years in this field, he now works with CMI-Canada as a speaker and writer. He is a contributor to *Evolution's Achilles' Heels.*

Donald Batten has a Ph.D. in plant and agricultural science. His work on the pollination of lychee and mango is internationally recognised. Managing Director of CMI-Australia and a prolific author, he is the contributing editor of *The Creation Answers Book.*

Peter Borger has a Ph.D. in molecular biology and is a freelance researcher, both in English and his native Dutch. He is the author of the in-depth book, *Darwin Revisited, or how to understand biology in the 21st century.*

Samuel Gan has a Ph.D. in allergy (molecular biology). Principle Investigator of A*STAR Singapore, he researches viruses, therapeutic antibodies, and psychology. He is founding Editor-in-Chief of the *Scientific Phone Apps and Mobile Devices* journal.

Introduction

By Philip Bell

This short book is a critical review of claims made by an expert panel of evolutionary scientists in the *New Scientist* 'Instant Expert' book, *How Evolution Explains Everything About Life: From Darwin's brilliant idea to today's epic theory.*[1] Those are massive claims, but can they be demonstrated? Well, by the end of the book, the *New Scientist* team feels sufficiently emboldened to conclude:

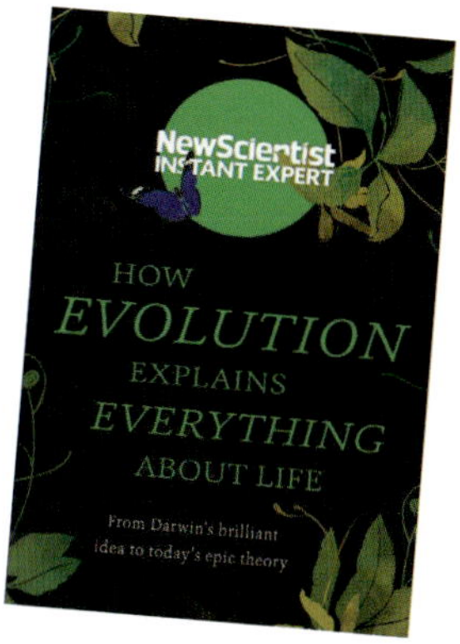

Fig. 1. Front cover of How Evolution Explains Everything About Life *(2017)*

> *"Innumerable examples of evolution in action can be seen all around us … Evolution is as firmly established a scientific fact as the roundness of the Earth" (p. 245).*

How often have you encountered such grandiose claims? The implication is that even to doubt big-picture evolution (molecules-to-man), certainly to critique evolution, is preposterous. It is as daft as insisting that the Earth is flat.

Imagine evolution as a grand pudding: the real proof of the pudding is in the eating. It is not sensible simply to rely on what the 'pudding experts' may tell you: 'Oh it really is the very finest of puddings, the combination of flavours and textures is simply superb. Cooked to perfection, it is quite mouth-watering. Once eaten, never forgotten. No bad aftertastes, easily digested, and reaches the parts that more inferior puddings cannot reach. Completely banishes hunger pangs, leaving you contented, without unpleasantly repeating on you later. Never did a more splendid pudding see the light of day!' Ah yes, but does it *really* live up to expectations? There's only one way to find out: eat some and test those mighty claims! Cut into it and sample the contents for yourself.

Similarly, to determine the sum and substance of *How Evolution Explains Everything About Life* (hereafter, HEEEAL), we must examine what the experts are saying. Without doubt, the book contains useful discussion on evolutionary topics, but can its lofty claims be substantiated? The very examination itself is an excellent opportunity both to showcase some fascinating scientific findings and, we are confident, to demonstrate the superiority of the case for supernatural creation.

Experts go head-to-head

This *New Scientist* 'Instant Expert' book represents the joint efforts of ten writers, the overall editor Dr Alison George (biochemist), and a panel of Ph.D. experts: Dr Susan Blackmore (psychologist), Dr Peter Bowler (science historian), Professors Adrian Bird (geneticist), Lee Alan Dugatkin (biologist), Steve Jones (geneticist), Kevin Laland (evolutionary biologist), George Turner (zoologist), and David Sloan Wilson (biologist and anthropologist), and Dr John van Wyhe (Darwin and Wallace historian).[2]

Consequently, it seemed fitting to assemble a panel of expert Ph.D.s (biblical creationists) to tackle the subjects raised—much as was done for the book and documentary project *Evolution's Achilles' Heels.*[3] A variety of audacious claims are made within the pages of HEEEAL, along with some telling admissions. Our own ten experts have responded in the very areas which are pertinent to their qualifications and professional experience:

- Dr Jonathan Sarfati (physical chemist)
- Dr Pierre Jerlström (molecular biologist)
- Dr Robert Carter (marine biologist and geneticist)
- Dr David Catchpoole (plant physiologist)
- Prof Stuart Burgess (mechanical engineer and biomimeticist)
- Dr Tasman Walker (mechanical engineer and geologist)
- Dr Jim Mason (nuclear physicist)
- Dr Donald Batten (plant and agricultural scientist)
- Dr Peter Borger (molecular biologist)
- Dr Samuel Gan (molecular biologist).

Not all of CMI's panel of experts read the book in full but all were provided with the chapter outlines and the full context pertaining to the quotations upon which they were invited to comment (shown hereafter in GREEN boxes). Therefore, in formulating their replies, each expert was fully cognizant of the scope and thrust of the writer(s) whom they were critiquing. In what follows, the headings and sub-headings are mine (not borrowed from HEEEAL) but are designed to capture the thrust of the quoted claims and admissions.

Audacious claims

Life's chemical origin was a cinch

A response by Dr Jonathan Sarfati

> *"The only thing we know for certain is that life must have popped into existence sometime between Earth's formation 4.5 billion years ago and the appearance of the first undisputed fossils, about 3.4 billion years ago" (p. 69).*

The above presupposes something they most certainly don't know for certain: that life evolved from non-living chemicals. This claim is called chemical *evolution*. However, it goes against what we do know about chemistry and information theory. So, arguments against chemical evolution are not appeals to ignorance, but about what really happens in chemistry.

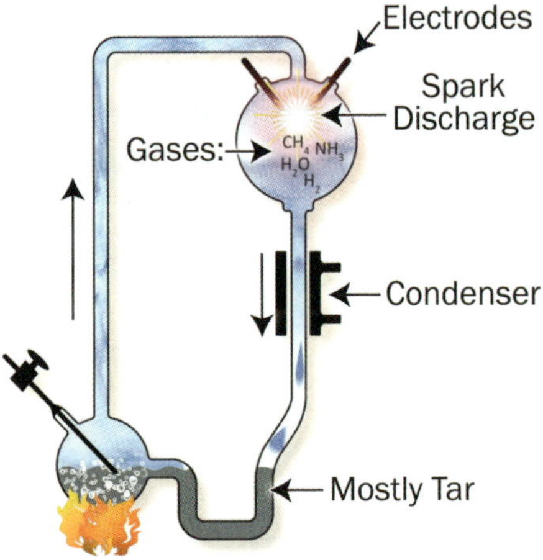

Fig. 2. Spark discharge apparatus used in the famous Miller-Urey experiments of the 1950s, designed to simulate chemical reactions on the putative 'early Earth'.

Many claims about chemical evolution sound impressive on the surface, but not so much after looking at the chemistry involved. Chemical evolution ('abiogenesis') is basically propaganda and includes the following:

1 Find a trace of compound A in a spark discharge experiment, and compound B in another simulation (sometimes with mutually incompatible conditions), then claim, "See, A and B can be produced under realistic primitive-earth conditions."

2 Obtain pure, homochiral,[4] concentrated A and B from an industrial synthetic chemicals company, react them to form traces of the more complex compound C. Then the news is trumpeted that C will form under primitive earth conditions.

3 Presume that dilute A and B can react that way, and that they won't react destructively with contaminants D, E, or F that were also formed in the first experiments.

4 Hope people don't notice that C didn't form abiotically (in the absence of life) but needed much intelligent interference.

5 Now that C has been 'proven' to form abiotically, chemical evolutionists purchase pure, homochiral C, and find ways of reacting it to produce traces of biopolymer X among many other things (biopolymer = many small biomolecules bonded together).

6 Purchase pure homochiral biopolymer X, and repeat these steps.

But any synthetic chemist knows that lots of intelligent work must go into a multi-step process. Real chemistry is not very cooperative.

> *"Life, it seems, is a matter of basic chemistry—no magic required, no rare ingredients, no bolt from the blue. And that suggests an even more intriguing possibility ... life may have had many origins" (p. 75).*

Not so:

Information is a huge stumbling block for chemical evolution

Much of this is only the building blocks, and ignores the vital concept of *coded information* or *specified complexity*—how they are put together. The flaws can be illustrated by exactly analogous claims:

1. We have just mixed flour and water and applied heat, and formed the letters to make alphabet soup (i.e. assume for a moment that the building blocks of life really have been produced).
2. Sometimes a few letters can stick together (i.e. assume that the building blocks can combine to some extent).
3. Therefore we showed that the works of Shakespeare could also have formed naturalistically.

We can see that the works of Shakespeare are more than the component letters. We can't explain them by looking at the chemical properties of the alphabet soup, or of the ink molecules on a printed page, or the computer memory. Something clearly distinguishes the works of Shakespeare from a random agglomeration of alphabet soup letters. This something is information. Information is not matter or energy. See for yourself: arrange alphabet soup letters into a Shakespearean sonnet on a plate, weigh them all. Then scramble the letters, thus destroying the information. Weigh it again. No change! So, information has no weight, therefore no mass.

Living cells require information, and decoding

Even the simplest living cell has an enormous quantity of information on its DNA (deoxyribonucleic acid)—about 600,000 'letters'. And it has the decoding machinery to 'read' this information. But the instructions to build this decoding machinery are encoded on its DNA. DNA can't be decoded without

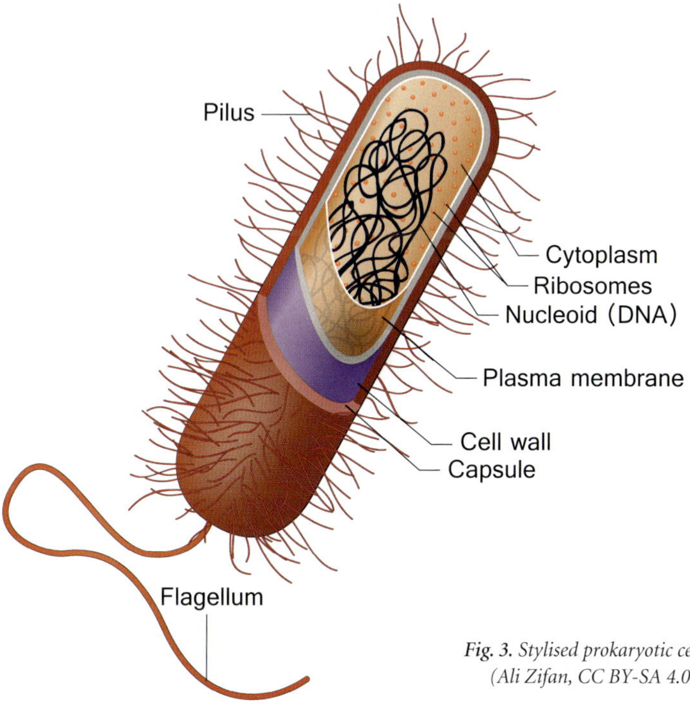

Fig. 3. Stylised prokaryotic cell (Ali Zifan, CC BY-SA 4.0).

machines, but the machines can't be built without the DNA coding for them. This is a vicious circle for chemical evolution.

Also, the machines must be extremely accurate right from the start. If not, then coding errors will creep in. Then the next generation of decoding machines will also have bugs, and they will do an even worse job of decoding, making even more bug-infested machines, which do a still worse job of decoding … This is called *error catastrophe,* and will result in extinction.

In general, if we have a genome of *n* letters, then the number of mistakes must be no more than 1/n. E.g. if it were 100 letters, then only a 1% error rate could be tolerated; 1,000 letters could tolerate 0.1%. Living cells have error-checking machinery, but they are encoded on 1,000 letters or more. This leads to the 'Eigen paradox': longer DNA polymers are more vulnerable to errors, leading to error catastrophe. But it needs a long polymer to code for any repair enzyme!

Science should be about the most logical explanation, not necessarily restricted to a materialistic explanation. The logical explanation for the above is that life could not have started imperfectly and improved. Rather, it must have started practically perfect (functional), otherwise it would not work at all. This is consistent with Genesis 1–3: God created everything "very good", and because of Adam's sin, things have been deteriorating ever since. Therefore, it is simply false to claim that life could have originated many times because it "is a matter of basic chemistry".

RNA world

> *"[Research] findings suggest that building a rudimentary RNA world may not have been the special, once-in-a-universe occurrence it is popularly made out to be. This raises an intriguing possibility: that life's earliest stages didn't happen just once, but over and over again. If this is true, then life's first epoch*

> *was one of great experimentation. Many different kinds of live molecular machines would have popped up in the primordial soup, some more successful than others" (pp. 80–81).*

The RNA world was proposed to try to answer the vicious circle described above. Its proponents hope that one molecule can perform the tasks of both coding and function. Modern cells have DNA and RNA (ribonucleic acid) for coding, and proteins for function.

A huge problem is that RNA is very unstable. DNA is not as unstable, because as the name suggests, an oxygen has been removed. That is, a hydroxy group[5] has been replaced by a hydrogen. This hydroxy group of RNA is easily attacked.

But DNA is also unstable. The 2015 Nobel Prize for Chemistry was awarded to three researchers for the discovery of the instability of DNA, which led scientists to realize that living things must have

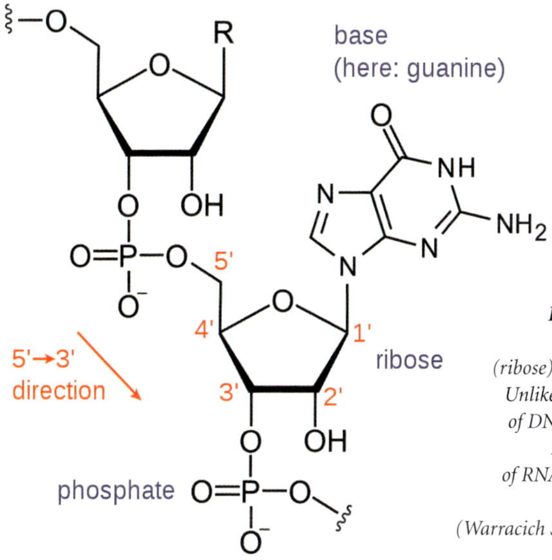

Fig. 4. The backbone of RNA is made of sugar (ribose) and phosphate (PO^4). Unlike the deoxyribose sugar of DNA (which has H at the 2'-position), the ribose of RNA has a hydroxyl (OH) group at this position (Warracich Sahib, CC BY-SA 3.0).

repair machinery. Even under favourable conditions—protected deep inside bones, and frozen at –5°C, it would be completely fragmented in 6.8 million years.[6] At higher temperatures DNA breaks down even more quickly. At the temperatures of hydrothermal vents, proposed by some as the beginning for life—forget about it! Since DNA is too unstable to persist for long, *a fortiori,* the idea of an RNA world is even more preposterous.

This is a good time to point out that chemical evolutionists often appeal to vast time periods. But deep time would be the enemy—more time for everything to break down.

Furthermore, even the component building blocks of RNA are unstable. Take ribose for example, the sugar in the RNA backbone. The usual explanation is that this originated by the polymerization of formaldehyde in an alkaline solution, called the formose or Butlerov reaction. But one problem is that this is one of the huge jumble of products produced, so we are back to asking how ribose was purified. And there are also several ways that ribose can be arranged (called isomers), and only one is suitable for RNA. An even worse problem is that the very same alkaline conditions quickly destroy aldose sugars—including ribose—via the Cannizzaro reaction. This problem is worse still under high temperatures, such as the alkaline hydrothermal vents.

RNA also has phosphate in its backbone. But where would that come from? Phosphate readily combines with metal ions in seawater to form an insoluble precipitate, so would be unavailable.

Then somehow we need to get ribose, phosphate, and the RNA 'letters' to combine into the building blocks, called ribonucleotides. But these chemicals don't readily combine at all, let alone in just the right way to form ribonucleotides. Then somehow the ribonucleotides must combine, although in water the reaction tends to go in the opposite direction. Then we have the problem mentioned above: ribonucleotides are like alphabet soup, and RNA is like the works of Shakespeare.

Many claims about RNA's ability have been highly overstated in RNA world scenarios. We have no actual examples of a genuinely self-replicating RNA. The nearest is an engineered ribozyme 202 nucleotides (nt) long called tC9Y that could make an RNA strand 206 nt long from activated nt building blocks, with a maximum fidelity of 97.4%. This sounds high, but as explained earlier, to avoid error catastrophe, the copying fidelity must be 1 in 202, or 99.5%. Other experiments are not really replication but *ligation*—one RNA strand joining two halves that must have a matching sequence. Of course, there is no explanation of how a primordial soup could generate these matching RNA strands.

Although RNA can supposedly both reproduce and catalyze, for most applications, *a given RNA molecule* can't do both. Most enzymes need a 3D structure, which in turn needs the RNA molecule to bond with itself. But to reproduce, the RNA chain must be open to bond with new RNA building blocks.

In general, the RNA world has far too many problems to be a viable chemical evolutionary theory. It's no wonder that biochemist Harold Bernhardt, working in New Zealand, wrote a paper, "The RNA world hypothesis: the worst theory of the early evolution of life (except for all the others)".[7] Of course, he means, worse than any others that are *materialistic.*

Parasitic DNA equals stupid design

A response by Dr Pierre Jerlström

> *"And as sequence data continues to pour in, biologists are well on their way to working out how every one of our 20,000 or so genes are involved" (p. 55).*
>
> *"...the genomes of complex creatures reveal a lack of any intelligence and foresight. Your DNA consists largely of millions of defunct copies of parasitic DNA. The inescapable conclusion is that if life was designed, the designer was lazy, stupid and cruel" (p. 124).*

The number of protein-coding genes believed to be contained in the human genome has been revised several times in the last few decades: from 60,000–80,000 in 1995, to 140,000 genes in 1999, and to 19,000–20,000 in 2003, as determined from the Human Genome Project. However, the current figure of 20,000 to 25,000 protein-coding genes only represents about 1% of the genome.[8] Consequently, the original interest in coding genes has now shifted to uncovering possible roles for the remaining 99% of the genome.

Studies of the non-coding part of the human genome, making up most of its three billion nucleotides, initially revealed many simple, repeated sequences with no apparent function. It was postulated that these sequences are a result of mobile DNA molecules that just replicate and spread copies of themselves within the host genome. This idea gave rise to the selfish/parasitic DNA hypothesis. Repeated sequences resembling leftovers of viruses were also found, as well as sequences named 'pseudogenes' which appeared to be non-functional genes that had lost their ability to produce any proteins. All these non-protein-coding types of DNA are now commonly referred to as junk or vestigial

DNA, based on the belief that they are useless leftovers from past evolutionary events.

Evolution and the need for junk DNA

The junk status of non-coding DNA has been embraced to support neo-Darwinian evolution without experimental support for several reasons:

- Junk DNA presents a huge store of spare/available DNA that, it is claimed, could randomly mutate into new genes advantageous to the cell.

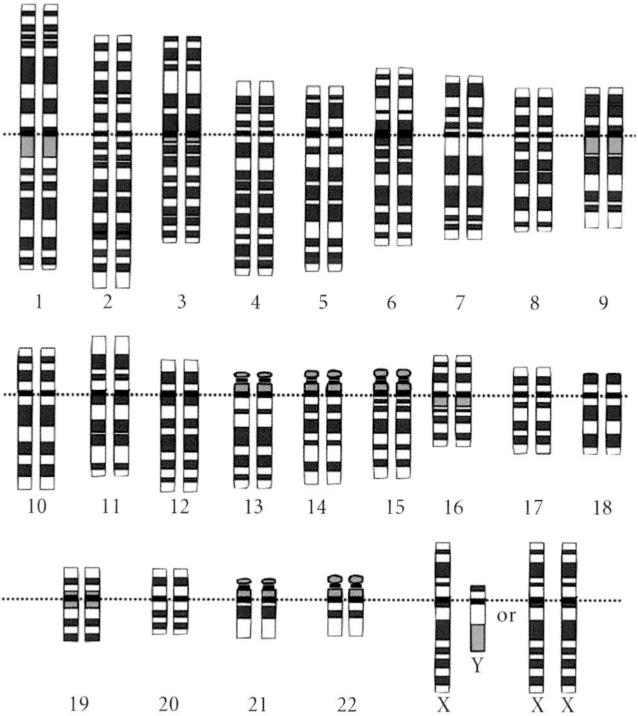

Fig. 5. The typical diploid human karyotype of 23 pairs of chromosomes: 22 pairs of autosomes (non-sex chromosomes) plus a pair of sex chromosomes, either XX or XY (National Human Genome Research Institute, NIH).

- It is also needed to solve mathematical problems: if most of the DNA were functional, then there is too much variation, too much DNA to mutate, and too few generations to validate the theory of evolution, even with deep time. A good example is the supposed divergence of humans and chimps from a common evolutionary ancestor around 6 or 7 million years ago.
- The idea of junk DNA is also required to solve the lack of correlation between the genome size of different eukaryotes, and their complexity and number of protein-coding genes. For instance, humans have approximately the same number of genes as less complex organisms such as sea anemones and jellyfish.
- In contrast to the standard neo-Darwinian Model, the Neutral Model of evolution is not selection driven, but evolutionary changes are instead a result of random mutational changes that result from genetic drift. However, this model also depends on most DNA in genomes being non-functional.

Surprising designed functions of junk DNA

However, experimental science has been steadily revealing the important roles played by many DNA sequences formerly lumped into the junk pile. These include genome structure and function, gene regulation, and even rapid speciation.

Moreover, the finding of non-protein coding genes that express regulatory RNAs has now raised the total number of 'genes' in the human genome to approximately 47,000, plus about 2,300 micro-RNA genes. Additional functional DNA that encodes neither RNA nor proteins has also been discovered.

The story of junk DNA is reminiscent of the 'vestigial organs' claim, i.e. organs believed to be non-functional remnants from our evolutionary past. Functions have now been found for virtually all of the nearly 200 organs that were once claimed to be useless/junk. We now know that the main reason for labelling these organs as such was both a lack of understanding of their function, and a mindset locked into an evolutionary worldview.

According to genetics Prof John Mattick, "the failure to recognize the full implications of [the non-coding DNA] may well go down as one of the biggest mistakes in the history of molecular biology".[9]

Fig. 6. Illustration by Sarah Fretwell.

Contrary to claims that "genomes reveal a lack of any intelligence and foresight" and that "the designer was lazy, stupid and cruel", ongoing research is elucidating the structure and the many functions of DNA elements in our genome. Especially surprising has been the wide variety of non-protein-coding DNA sequences once believed to be mere junk. These include:

- Pseudogenes: genes that look like they have lost their ability to produce any proteins. They can switch off genes when they are not needed, or on when more proteins are required.
- Retrotransposons: mobile genetic elements that jump from one place in the genome to another. They make up a large part of the human genome and enable genes to be 'read' in the opposite direction. They are particularly active during embryo development, suggesting an important role in gene regulation associated with development.
- Introns: non-coding DNA regions located within protein-coding genes which are removed before a protein is made. They appear to stabilize closed chromatin structures in, and around, genes, and the DNA sequences involved in controlling them. Control sequences have been found, contained within introns.
- Telomeres: repeated sequences at the ends of chromosomes that protect them from being degraded during the copying of genetic material.
- Stretches of 'junk DNA' can also act in a mechanical manner like that of 'scaffolding', instead of being 'read' to carry out a function.

The ENCODE (Encyclopedia of DNA Elements) project revealed something unexpected: that over 80% of the human DNA is biochemically active.[10] The bulk of the genome, then, made up of non-protein coding sequences, appears to be involved in controlling the proteins produced from only a small part of it (about 1%), the coding DNA. This emphasizes the essential nature of noncoding DNA, and how it is far more important than originally believed.

Even the way the coding part of the genome functions is much more intricate than originally thought. We now know that a single gene can produce many different proteins, via a process called "alternate splicing", where exons (the protein-coding portions in the genome) from different locations are joined to create complex combinations. Moreover, numerous overlapping RNAs ('transcribed' from DNA) are often produced from the same stretch of DNA.

As our understanding of our genome has steadily increased, it has become more obvious how amazingly and carefully designed the human genome is. Instead of a so called 'hodgepodge', our DNA is the most complex, ingenious and awe-inspiring information system that mankind has ever witnessed. According to Bill Gates, co-founder of Microsoft: "DNA is like a computer program but far, far more advanced than any software ever created."[11]

The variety of non-coding DNAs and their many functions shows that the idea of our genome being full of junk from past evolutionary events is an argument from ignorance, and one that has hampered the progress of scientific knowledge.

The ENCODE project that finally destroyed the idea of 'junk DNA' was given wide publicity in 2012—why is *New Scientist* continuing to state nonsense about 'junk DNA'? How up-to-date are the 'experts' who were consulted?

New genes evolve through lucky mutations of junk DNA

A response by Dr Robert Carter

> *"Most mutations make no difference because they occur in junk DNA, which makes up the vast majority of the genome" (p. 130).*

In order to crack open the 'junk DNA' claims, I would like to use an example of what was thought to be unnecessary information in the genome: the third letter of the codon. Each amino acid in a protein is coded by three consecutive DNA letters. But since there are 64 possible three-letter codes and only 20 amino acids, there are a lot of duplications (see fig. 7 for an equivalent RNA codon table). But the code is not random. For example, CGA, CGC, CGG, and CGT all code for the amino acid arginine. Any mutation to that third letter still codes for arginine. The third

1st base	2nd base								3rd base
	U		C		A		G		
U	UUU	(Phe/F) Phenylalanine	UCU	(Ser/S) Serine	UAU	(Tyr/Y) Tyrosine	UGU	(Cys/C) Cysteine	U
	UUC		UCC		UAC		UGC		C
	UUA	(Leu/L) Leucine	UCA		UAA	Stop (*Ochre*)[B]	UGA	Stop (*Opal*)[B]	A
	UUG[A]		UCG		UAG	Stop (*Amber*)[B]	UGG	(Trp/W) Tryptophan	G
C	CUU		CCU	(Pro/P) Proline	CAU	(His/H) Histidine	CGU	(Arg/R) Arginine	U
	CUC		CCC		CAC		CGC		C
	CUA		CCA		CAA	(Gln/Q) Glutamine	CGA		A
	CUG[A]		CCG		CAG		CGG		G
A	AUU	(Ile/I) Isoleucine	ACU	(Thr/T) Threonine	AAU	(Asn/N) Asparagine	AGU	(Ser/S) Serine	U
	AUC		ACC		AAC		AGC		C
	AUA		ACA		AAA	(Lys/K) Lysine	AGA	(Arg/R) Arginine	A
	AUG[A]	(Met/M) Methionine	ACG		AAG		AGG		G
G	GUU	(Val/V) Valine	GCU	(Ala/A) Alanine	GAU	(Asp/D) Aspartic acid	GGU	(Gly/G) Glycine	U
	GUC		GCC		GAC		GGC		C
	GUA		GCA		GAA	(Glu/E) Glutamic acid	GGA		A
	GUG		GCG		GAG		GGG		G

Fig. 7. RNA codon table for the standard genetic code. Note: in RNA, uracil (U) replaces the thymine (T) of the DNA. Colours indicate the properties of the coded amino acids: yellow = nonpolar, green = polar, blue = basic, pink = acidic; grey = a stop codon (Wikipedia).

letter is called the 'wobble base' and we see a lot of mutations in those positions.

We learned years ago that genes that are used together tend to have similar codon preferences. In one set of genes, maybe most of the arginines are coded by CGA, but in another set of genes maybe most of the arginines are coded by CGT. These genes tend to group with each other, in 3D space, right next to a nuclear pore, so the set of mRNAs can be translated as a packet.[12] Later, we learned that there are different classes of ribosomes that are fine-tuned to work with different sets of genes, based on their codon preferences. More recently, we learned that changing the wobble base tells the ribosome 'this mRNA does not belong here' and this mRNA is then marked for quick degradation.[13] Codon degeneration (due to mutations) has even been linked to the loss of viability of the human H1N1 influenza virus.[14] And we now have multiple cases where a protein can be ruined by changing a codon, *even if it codes for the same amino acid*.[15] This last part was a surprise because everybody thought the wobble base was

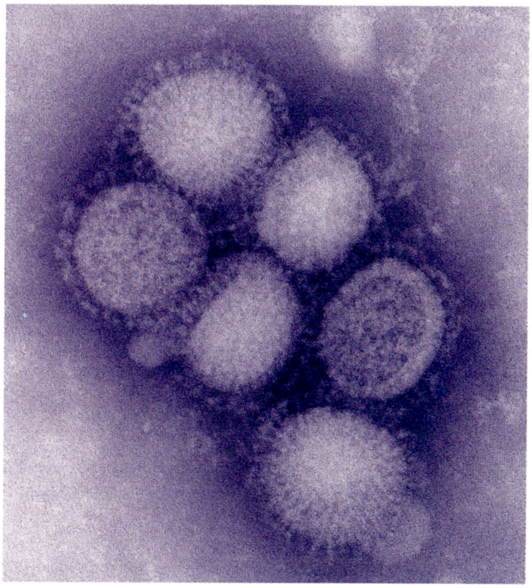

Fig. 8. *False colour electron micrograph of the H1N1 influenza virus, responsible for both the 1918 flu and 2009 swine flu pandemics.*

not important. Instead, what we have learned is that by replacing a common codon with a rare codon, the translation machinery might slow down or pause as it searches for the right transfer RNA. But while things are slowing down, the protein might be folding (proteins must fold to produce their functional 3-D structure). And that pause might cause a critical misfold. In other words, the *speed* of translation is intrinsic to protein folding and the speed can be regulated by codon use.

Thus, one of the best cases for mutational robustness (that a lot of mutations don't matter) has bitten the dust. You can't just mutate the genome willy-nilly. The cell is a fine-tuned machine, with multiple levels of error checking and optimization. True, it can withstand a considerable amount of mutation. But is this because most of the information content is just 'junk'? Or is it because it is a marvellously engineered system, with inbuilt redundancy and error checking, and high levels of fault tolerance?

The more we learned about the genome, the more we realized the latter was the correct answer. The 'junk DNA' argument is nothing more than the vestigial organ argument of days gone by. No, the embryo does *not* go through the stages of evolution as it develops[16] and, *no*, the human body does not have vestigial, leftover, or reduced organs that are no longer needed.[17] These were 'arguments from silence' and were based on our lack of knowledge. Thus, all we have to do is find uses for the 'junk' DNA and the argument dies. Yet, the list of *functional* 'junk' DNA keeps getting longer and longer.[18]

> *"...even though at least half of our genome is junk, as much as 90 per cent of it can be accidentally transcribed into RNA on occasion. This means that it might not be that uncommon for random bits of junk DNA to get translated into protein.*

> *Since most random proteins will probably be harmful, natural selection will eliminate these DNA sequences, but occasionally one will strike it lucky. A sequence that does something beneficial will spread through a population and rapidly evolve into a new gene, becoming optimized for whatever role it plays" (p. 55).*

I note that the estimates of the amount of junk DNA have been drastically revised downward in recent years. It used to be closer to 98%. This book is now claiming 50%. What happened? We learned about promiscuous RNA generation from what we thought were inactive portions of the genome. In fact, most of the computations happening in the cell are at the level of RNA, not protein. There are innumerable decisions being made all the time, among RNAs, which then control which genes are turned into protein. In essence, we are RNA computers. That is, you can think of the DNA as your computer ROM, the RNA as the RAM, and the proteins as the output.[19] In a computer, all the calculations are done at the level of the volatile system memory, the RAM.

Evolutionists have struggled to explain the source of new genetic instructions. An old idea is 'gene duplication'.[20] It was thought that the two copies of an accidentally duplicated gene could evolve along separate pathways. While it might be possible to explain the evolution of genes within the same family and with similar functions

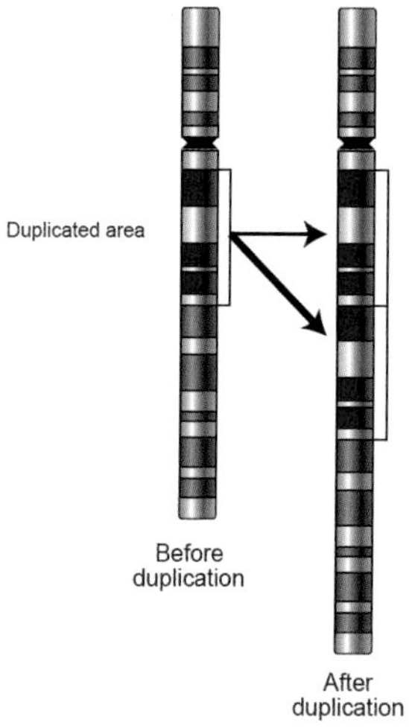

Fig. 9. Duplication of a region of a chromosome, containing one or more genes (Courtesy: National Human Genome Research Institute).

(e.g. Hox gene family that regulates aspects of embryo body plans), this cannot explain the origin of radically different genes. The idea being presented here involves something I call 'genetic co-option'. That is, a random string of letters, if it accidentally translated to protein, might do something, "strike it lucky". Yet, the authors provide no details, or examples. How much of the genome must be turned on at random with each generation to allow for enough experiments to occur? How much parasitic DNA must exist? What is the actual probability that random DNA will have a function?[21] How much time would it take? Simulations show that even with millions of years nothing significant can happen.[22]

They do not bother to answer such questions. They provide no diagrams and no details, just broad brush statements that appear to give a semblance of science to the bald assertions that follow. This is par for the course. Darwinists have been doing this since 1859, when Charles Darwin published a book with much theory but no substance.

Darwin's 'Tree of Life' is still very useful

A response by Dr David Catchpoole

> *"The tree concept has been the unifying principle for understanding the history of life on Earth. Darwin assumed that descent was exclusively 'vertical', with organisms passing traits down to their offspring. But what if species also routinely swapped genetic material with other species, or hybridized with them? … We now know that this is exactly what happens, mainly in the realm of single-celled organisms. … The number of horizontal transfers in animals is not as high as in microbes, but can be evolutionary [sic] significant. Even so, nobody is arguing—yet—that the tree concept has outlived its usefulness in animals and plants" (pp. 83–86).*

On the contrary, it is pertinent to argue for:

A reality-check versus creative imagination

The evolutionary concept of a 'tree of life' has indeed been a unifying principle for understanding a supposed evolutionary 'history' of life on Earth. A measure of its importance to evolutionary theory is the fact that when Charles Darwin introduced the idea in his *Origin of Species*, his tree-of-life sketch was the book's *one-and-only* illustration (see fig. 10). Evolutionary biologists today have gone so far as to say that the tree-of-life concept was so absolutely central to Darwin's thinking that without it evolutionary theory would never have happened.[23]

However, the concept fails a reality check.

Since the time of Darwin, the tree has been continually subject to change, in an effort to make it fit contradictory biological *evidence*.

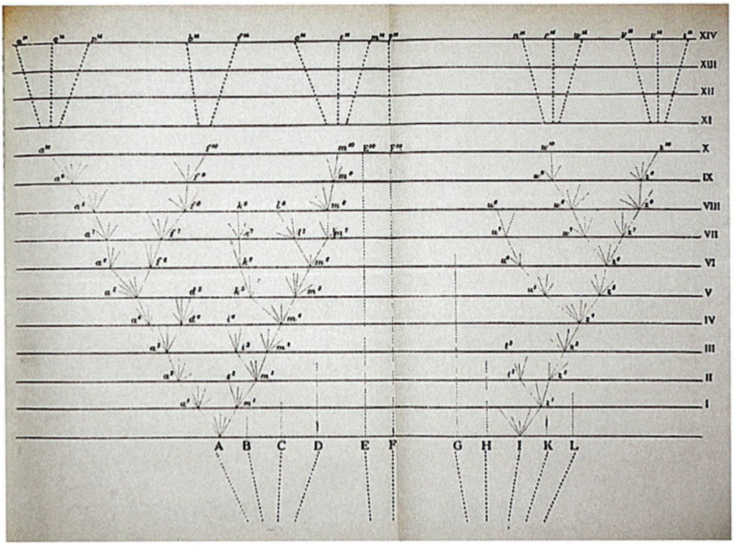

Fig. 10. Darwin's tree diagram, the only illustration in the first edition of The Origin of Species by Means of Natural Selection, or the Preservation of Favoured Races in the Struggle for Life (1859).

Initially, an organism's position on the tree was determined by its perceived complexity ('simpler' organisms being placed closer to the tree's 'root') and its presumed anatomical/morphological similarity to other species (placing them on the same or nearby 'branches').

But molecular/genetic studies have progressively revealed massive differences between organisms that previously were presumed to be similar. As our genetic knowledge has increased, the large gaps between many supposed relatives have presented an ongoing headache for evolutionists. 'Rogue' genes plague the tree-of-life concept; e.g. genes restricted to certain organisms, genes that 'appear' without any evolutionary precursors, genes that produce completely different 'trees' to other genes, and mosaics of DNA sequence patterns that cannot be explained by common ancestry.[24]

And some organisms just do not fit anywhere on the tree at all. In relation to Pandoraviruses (discovered in 2013), evolutionary researchers wrote, "Because more than 93% of Pandoraviruses genes resemble nothing known, their origin cannot be traced back to any known cellular lineage." Hence they proposed that Pandoraviruses came from a "different tree of life altogether"![25]

As for the 'root' of the (main?) tree of life (fig. 11A), evolutionists have variously suggested many candidate organisms, but none fit the bill, especially with our increasing genetic knowledge. E.g. a DNA analysis of 29 animal species "calls into question the very root of the animal tree of life".[26] That analysis placed the fragile comb jellyfish, which has well-developed tissues, *earlier* in time than

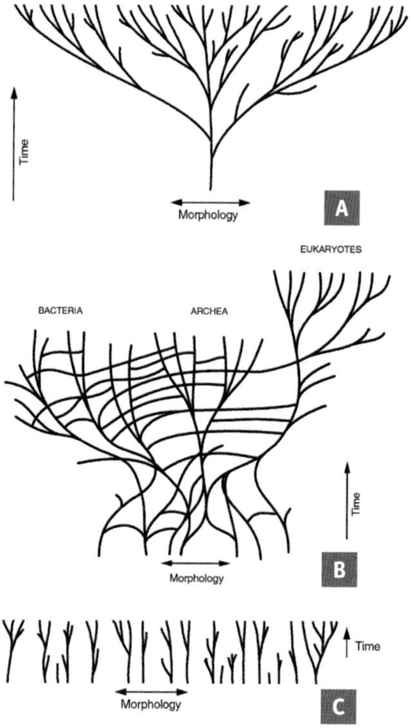

Fig. 11. Different trees representing contrary ideas of phylogeny. A) Standard evolutionary tree—views all species as having descended from a common ancestor. B) Revised evolutionary tree—really a tangled bush, depicting gene horizontal transfer among branches and a community of ancestral cells. C) The 'creationist orchard'—diversity has occurred since the creation of the original Genesis 'kinds' (baramins).

the sponge, "which has no tissue to speak of".[26] That is the exact reverse of both Darwin's tree concept and a fundamental tenet of evolutionary theory, which requires the simple to evolve first, then the more complex.

Also, the widespread awareness now that single-celled (and sometimes other) species swap genetic material has led some to dub the tree a 'web', or a 'tangled bush' (see fig. 11B). None of this sits well with the *New Scientist* book's title, *How Evolution Explains Everything About Life*! An earlier article in *New Scientist* magazine itself didn't mince words. Under the heading "Why Darwin was wrong about the tree of life", the text further declared: "The tree of life, one of the iconic concepts of evolution, has turned out to be a figment of our imagination."[23]

As Eric Bapteste, evolutionary biologist at the Pièrre and Marie Curie University in Paris, France, quipped, "We have no evidence at all that the tree of life is a reality."[23]

Exactly right. We see that the evolutionary 'tree of life' has no roots, no trunk, no boughs, or medium-sized branches. Those were all figments of imagination. The only reality are the living leaves and twigs at the tips—what have been referred to as the 'creationist orchard'—matching the evidence of built-in variation within the various biblical kinds (see fig. 11C).[27]

However, *New Scientist* and the wider evolutionary fraternity don't want to surrender the tree-of-life concept, irrespective of the evidence against it. As Eric Bapteste mused, "If you don't have a tree of life, what does it mean for evolutionary biology?"[23] In other words, if it's goodbye to the evolutionary tree of life, it's goodbye to evolutionary theory too. No wonder *New Scientist* magazine doggedly said that "downgrading the tree of life doesn't mean that the theory of evolution is wrong ... it's just that it is more complex than Darwin imagined."[23]

More complex than Darwin imagined? That requires a strong imagination!

So it comes down to this: *Reality*, or *imagination*. It's a clear-cut choice.

Blind evolution innovates and creates

A response by Prof Stuart Burgess

> *"Evolution's methods are blind, brutish and aimless, yet it has furnished some of the most exquisite machines in the known universe. And every now and then, it stumbles across a truly stunning innovation that rewrites the rules of life" (p. 95).*
>
> *"The remarkable diversity of life on Earth stands as a grand testimony to the creativity of evolution. Over the course of 500 million years, natural selection has fashioned wings for flight, fins for swimming, and legs for walking, and that's just among the vertebrates. The capacity for evolutionary innovation—or in buzzword form, 'evolvability'—is built into the fabric of life" (p. 151).*

If evolution were a scientific fact, then it would follow that it *has* made stunning innovations. However, evolution is not a scientific fact and, therefore, the claim that evolution is able to innovate requires justification and explanation. This book provides no such justification or explanation.

It is revealing that the author admits that "evolution's methods are blind, brutish and aimless". When engineers develop methods of innovation, they avoid any methods that are "blind, brutish or aimless" because such methods do not work. Conceptual design is like searching for a needle in a haystack because innovative designs require unique solutions and unique combinations of physical effects. When you are searching for a needle in a haystack, you can only make progress if you are searching in an intelligent, goal-directed way.

In order to produce innovation, there are generally two things that are required. The first is the generation of a range of innovative concepts. The second is the selection of the best concept. Of these two things, it is the generation of concepts that is the really hard part. Selecting the best concept is important but relatively straightforward. I have seven registered patents for unique types of mechanical devices and in each case the generation of the concept was a very difficult task requiring the simultaneous bringing together of many aspects of technology.

It is revealing that the author focuses on the power of 'natural selection' and gives little or no detail on the generation of new concepts. While natural selection is a clear scientific fact, it cannot by itself innovate anything. A process of selection itself has no power of innovation at all because, by definition, it can only select from what already exists.

The evolutionary process of step-by-step change is inherently limited in what it can generate. This has been admitted by Steven Vogel who was a leading biologist at Duke University, USA. He said, "The evolutionary process faces constraints far more severe than anything impeding human designers."[28] The clear consequence of that statement is that evolution must be far worse at innovation than any engineer who is not so limited to step-by-step change. When I teach undergraduate engineering students how to innovate, I advise them to avoid step-by-step change because it won't get them anywhere!

Fig. 12. Orville Wright piloting the 'Wright Flyer' on 17 December 1903, with his brother Wilbur running alongside (John T. Daniels, 1873-1948). Their design brilliance was unimpeded by the unsurmountable constraints that face a blind evolutionary process.

The Wright brothers who invented the first controlled aircraft at the beginning of the twentieth century were good examples of great innovators. They brought together several key pieces of technology to make one giant leap of innovation. For example, they brought together control flaps, aerofoil-shaped wings, super-lightweight structures, and a power unit. The reason they could invent the aircraft was, firstly, because they understood the physics of flight (aerodynamics) and, secondly, that they realized how to bring together just the right technology to exploit specific physical effects. In contrast, incremental evolution, by its very nature, is incapable of bringing together such complex technology blindly and in one step.

If evolution were such an effective form of innovation, then engineers would want to copy it. However, engineers have not copied the principle of evolution. Computer programming makes it possible to speed up the hypothetical evolutionary process in order to 'visualize' it over very short time scales. Engineers have tried to simulate evolution in the form of 'genetic algorithms'. They have found that genetic algorithms can indeed be very good at fine-tuning but are always very ineffective at innovating. Being good at fine-tuning is not surprising because adaptation (and breeding) is well known to be an effective way of optimizing a design. But adaptation is not an example of evolution because it does not involve the generation of new codified information or novel design features. The failure of genetic algorithms to provide effective innovation is strong evidence that evolution is incapable of innovation.

The audacious claim by the author that evolution is capable of "stunning innovation" (so called 'evolvability') is simply based on the assumption that all organisms have evolved, not on any rational, evidence-based argument—it is an entirely circular argument. I would argue that evolution is stunningly ineffective at innovation!

No fossils have been found that challenge evolution

A response by Dr Tas Walker

> *"All it would take is one or two exceptions to challenge the theory. If the first fossil amphibians were older than the first fossil fish, for example, it would show that amphibians could not have evolved from fish. No such exceptions have ever been found anywhere" (p. 123).*

This statement repeats a common but thoroughly fake assertion about fossils and evolution. It is designed to lead people to think that evolution is supported by the fossil evidence, and that if the evidence were different, then evolution would be challenged. However, it does not work this way. No matter what is found, evolution would not be challenged because evolution is already accepted as true, *a priori*.

The reality is that fossils are being found in exceptional places all the time, but this does not challenge evolution. All that happens is that evolutionists change their story about how evolution works. And there are many ways the story can be changed such that it still seems neat and tidy.

HEEEAL gives a hypothetical example: "If the first fossil amphibians were older than the first fossil fish, for example, it would show that amphibians could not have evolved from fish." However, such a development would not cause a change in thinking like that. Evolutionists would have a multitude of ways to challenge many aspects of such a 'find'. Perhaps the 'amphibian' is not actually an amphibian. Perhaps the date assigned to the amphibian is too old, or that assigned to the fish is too young. Perhaps we have not yet found the earliest fish fossil. Or perhaps the amphibian and the fish evolved from a common ancestor that existed much earlier, of which we have not found fossil evidence yet. And so on. In the meantime, evolution would be retained, and the problem would be shelved and forgotten while waiting for an agreeable explanation.

Fig. 13. Fossil snake, Eupodophis descouensi *said to be 92 million years old (Didier Descouens, CC BY-SA 4.0).*

For example, in 1999 scientists reported well-preserved fossils of two different kinds of agnathan fish from China found in Lower Cambrian strata.[29,30] This was at least 20 Ma, possibly 50 Ma, earlier than previously thought, contradicting the neat story of how fish evolved. However, this exceptional discovery did not compel evolutionists to challenge evolution. They simply changed the evolution story.

Fossils are continually being found outside the time range of what was previously believed about evolution. Examples include the origin of snakes pushed back 70 Ma,[31] dinosaur origins pushed back some 20 Ma,[32] the origin of flowering plants pushed back over 100 Ma,[33] and the origin of multicellular life being pushed back 1.5 Ga.[34] None of this challenges evolution in any way because they just change their story.

So, evolution is the narrative used to explain the arrangement of fossils, with the story changing as required as new evidence is discovered. But there is another powerful narrative that explains the fossils. Fossils buried all over the Earth are expected from Noah's Flood, described in Genesis chapters 6–8 of the Bible. Instead of evolution over millions of years, the fossils record the order in which plants and animals were buried during the one-year Flood about 4,500 years ago. They were all living on the Earth already, at the time the Flood began. It is not hard to appreciate that the global Flood would produce a definite order for fossils. For example, the first sediments would have been deposited in the deep oceans and would have buried marine animals. Slow movers and bottom-dwellers such as the trilobites would have been entombed first, while fish and ammonites would have escaped until later.

> *"...Darwin predicted that transitional fossils would be discovered, and millions—trillions if you count microfossils—have been uncovered" (p. 129).*

This claim is not correct. It is true that Darwin said (in *Origin of Species*) that the absence of transitional fossils was "the most obvious and gravest objection which can be urged against my theory." He did not say it was just one or two transitional fossils that were absent but a "finely graduated organic chain." And so, he spent chapter IX, "On the imperfection of the geological record", explaining why such chains were missing, and why such chains were not expected to be found (thus trying to make his grand idea unfalsifiable).

Darwin argued that a chain would never be revealed because the "geologic record" itself was imperfect. That is because: sedimentation occurred on Earth in limited geographic places, the sedimentation rate was usually too slow to bury and preserve creatures, soft bodied creatures especially would disintegrate before burial, most sediments deposited over Earth history have not been preserved due to uplift and erosion, relationships between many creatures are not of the ancestor-to-descendent kind, and other reasons. Darwin, imagining vast eons of geologic time, summarized his thinking, "I look at the natural geological record, as a history of the world imperfectly kept, and written in a changing dialect; of this history we possess the last volume alone, relating only to two or three countries. Of this volume, only here and there a short chapter has been preserved; and of each page, only here and there a few lines."

Fig. 14. Beautifully preserved fossil jellyfish and sand ripples found in Wisconsin (USA) argue strongly for superfast burial (AP Photo / Reed Saxon).

However, fossil discoveries have revealed a different picture from what Darwin painted. Soft bodied animals, such as jellyfish, have been found fossilized, contrary to his expectation. Animals are frequently found well preserved with evidence of rapid death and burial, contrary to the idea of slow and gradual sedimentation, providing direct evidence against the eons of time. And rather than filling in the gaps between forms, fossil discoveries have tended to reinforce the "distinctness of specific forms, and their not being blended together by innumerable transitional links", as Darwin described the fossil record. This of course, is exactly what is expected from a biblical creation perspective.

Nevertheless, lots of fossils have been touted as transitional, and their publicity gives the lay person the impression that evolution is continually being confirmed. Following Darwin's lead, evolutionists do not regard this evidence as necessary but consider it a bonus. However, most of these claimed transitional fossils are disputed because they are actually just variations within the created kinds (e.g. two species of shellfish where the only difference is the direction of twist of the shell). They do not indicate any major transitions have occurred. Other claimed transitional fossils are disqualified as further discoveries are made.

For example, the fossil fish *Tiktaalik roseae* was claimed to be intermediate between fish and tetrapods. However, tracks of footprints found in a quarry in Poland, which researchers dated according to the evolutionary paradigm to some 18 Ma earlier, indicated that tetrapods existed before the supposed transitional form.

In another example, the fossil *Rodhocetus* was claimed to be intermediate between land animals and whales. Early drawings of the animal showed a whale-like fluke and flippers, even though fossil evidence for the fluke and flippers was missing. However, further fossil discoveries revealed that the creature had neither a fluke nor flippers, rather a tail, hind limbs, and fore limbs.[35] Evolutionists still portray *Rodhocetus* as an intermediate, although the evidence no longer supports this.

Fig. 15. Chicago Field Museum's rendition of Rodhocetus, claimed to have lived from 49–41 million years ago (photo by Keaton Halley)

So, the claims in HEEEAL are misleading. They give the impression that evolution is based on solid empirical evidence, and that, if the evidence was not the way it is, evolution would be abandoned. But the fact is that the idea of evolution is unfalsifiable. No matter what fossil evidence is discovered the evolution story can be modified to accommodate it. And the alleged thousands of so-called intermediate fossils that are referenced are just fossils. Evolution does not need these fossils because it is already 'known' to be true without them. But the fossils serve a useful purpose in that they provide a powerful impression to the lay person that evolution is being demonstrated to be true. But people are not aware that evolutionists do not consider that they need fossil evidence to establish evolution, or that each of the transitional fossil claims is contested by other evolutionists, or that later discoveries change the story about individual fossils and cause evolutionists to quietly put the discredited fossil claims aside.

Revealing admissions

Deep time is essential for evolution

A response by Dr Jim Mason

> *"Traditionally, people in Christian Europe had believed that the world was about 6000 years old. This view was guided by interpolations from the Bible, which itself gave no date for creation. Gradually such beliefs were modified by Christian thinkers based on new information about Earth, gleaned from the growth of mining and the development of geology. By the early nineteenth century it was widely understood that Earth could not be a few thousand years old, but must be inconceivably ancient" (p. 3).*

Determining the age of the Earth from geology used extrapolations backwards in time from observations made in the present, combined with unprovable assumptions about geological processes in the past. The observations in the present are the vertical sequence of rock layers, the thickness of these layers, and the extent of erosion visible in such places as the Grand Canyon. The geological processes involved are sedimentation, erosion, volcanism, earthquakes, and crustal plate tectonics. The unprovable assumptions are that the spatial and temporal frequency of these processes, as well as their intensities, have always been essentially the same as we observe them to be today.

Since the present-day, observable rates and intensities of all these things are very low, it led to the (incorrect) understanding that

Fig. 16. Sedimentary layers and erosion evident at the Hoodoos of Bryce Canyon National Park, Utah, USA (Luca Galuzzi - www.galuzzi.it, CC BY-SA 2.5).

sedimentary strata and igneous rock formations required vast ages for their formation; thus, "Earth could not be a few thousand years old, but must be inconceivably ancient".

However, ample evidence from present-day geological observations indicates that the assumptions regarding historical rates are invalid. The Mount St Helens eruption clearly demonstrated that thick layers of rock can be deposited (and canyons subsequently carved) in just a few hours.[36] Many polystrate fossilized trees testify to rapid deposition,[37] and even granitic plutons, long thought to take great ages to form, are now understood to have formed very quickly.[38]

> *"A young Earth would also be a problem for evolution, since evolution by natural selection requires vast stretches of time—'deep time'—as Darwin realised"* (p. 123).

This is quite true but there is ample evidence from various lines of investigation that indicates that the Earth is, in fact, quite 'young', with some of the more compelling ones being reports of unpermineralized biological tissue—especially DNA—in dinosaur fossils.[39] Nevertheless, reference is made here to 'deep time'. Evolutionary writers typically appeal to radiometric dating as having irrefutably established that the Earth is 4.54 ± 0.05 billion years old ($4.54 \times 10^9 \pm 1\%$ years), with the quoted uncertainty (error) supposedly indicating that this number is very precise.

Careful examination reveals that the quoted uncertainty does not represent accuracy, but rather is the expected variation in the result if the same experimental procedure was repeated many times. Accuracy is the difference between the measured result and the correct value. It is just assumed that radiometric dating is accurate. But is this justified?

"But examine everything carefully; hold fast to that which is good"

1 Thessalonians 5:21, NASB.

Radiometric ages, like those originally derived from geology (as discussed earlier), rely on several assumptions about geologic and physical processes in the past, none of which can be independently verified.[40] If these assumptions were all true and the technique performed as advertised, then one would expect that it would accurately determine the age of rocks of known age, those volcanic rocks *observed* to form in the present (i.e. where the age is defined to be the time since solidification). Also, one would expect that the ages determined for the same rock using different decay chains would be the same—at least within the experimental uncertainty. There are many examples of where this is not the case. For example: Mount St Helens,[41] Mount

Ngauruhoe,[42] and the Grand Canyon.[43] A reasonable conclusion is that the assumptions are not valid and radiometric dating techniques do not perform as advertised.

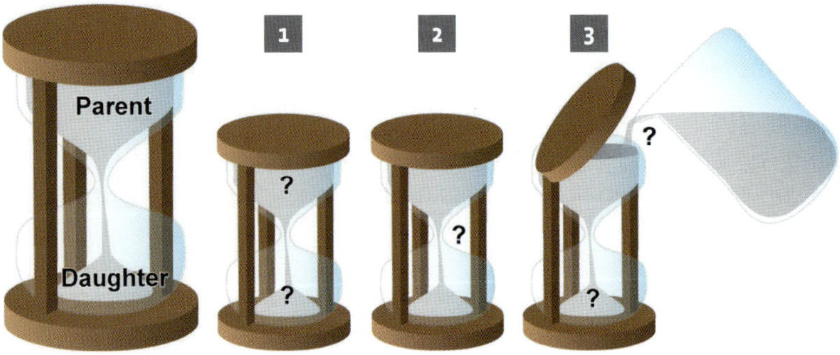

Fig. 17. Hourglass model to illustrate the assumptions involved in radiometric dating: 1) the initial amount of parent radioisotope and daughter decay product; 2) a constant decay rate; 3) a closed system—nothing added or removed from the sample since its formation.

The situation is further complicated by the fact that radiometric dating—with the debatable exception of radiocarbon dating—is not, and cannot, be calibrated. That is, there are no national standards, as there are for the kilogram or metre, against which the accuracy of the technique can be verified. Radiometric dating depends entirely on the validity of the underlying assumptions and evidence indicates that these assumptions are not valid.

Notwithstanding this, radiometric dating does provide evidence that does not rely on these assumptions, indicating that the Earth is indeed 'young' and "deep time" does not exist.

^{14}C in coal and dinosaur bones.

Fig. 18. Planet Earth (pexels.com, CC0).

If the entire world was composed of ^{14}C (radiocarbon), it would disappear entirely in less than one million years. Thus it is significant that ^{14}C has been found in coal that is allegedly between 30 and 300 million years old,[44] and dinosaur fossils that are allegedly more than 65 million years old.[45] This entirely contradicts the notion of "deep time". At most, these samples cannot be more than 90,000 years old (the design limit of modern accelerator mass spectrometers) and the measurements actually translate to an age of about 50,000 years using the standard approach. However, taking into account evidence reported in secular scientific journals, about the level of ^{14}C in the atmosphere when the rock layers in which the samples were found were deposited, the computed ^{14}C 'ages' are consistent with these coal and dinosaur samples having been buried about 4,500 years ago. Various schemes have been devised to try to explain the presence of this ^{14}C in a "deep time" scenario but all are cases of special pleading, and none survive detailed examination.

Radiogenic helium in zircons.[46]

Zircon crystals, when they form, incorporate ^{238}U but exclude ^{206}Pb. Thus, any ^{206}Pb in the crystals is considered to have formed from the radiometric decay of ^{238}U. For each ^{206}Pb atom formed, 8 alpha particles are emitted, which capture free electrons to become helium atoms. Helium , being a noble gas that does not form compounds, immediately starts diffusing out of the zircon crystals to regions of lower concentration. Over the billions of years, alleged to be the age of these crystals, essentially all of the

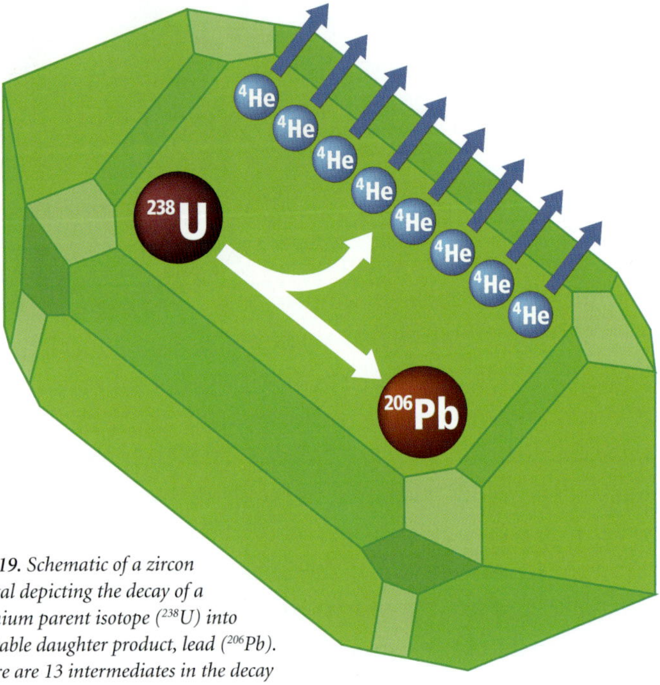

Fig. 19. Schematic of a zircon crystal depicting the decay of a uranium parent isotope (^{238}U) into its stable daughter product, lead (^{206}Pb). There are 13 intermediates in the decay series between uranium and lead (not shown).

helium would diffuse out. However, in some crystals, researchers have found that up to 50% of the helium that would have been produced was still present. Based on the measured amounts of helium, the rate of diffusion of helium through the zircons was calculated (as indicated by the ratio of Pb to U), firstly under the premise that they were 1.5 billion years old, and secondly, assuming that they were 6,000 years old (as would be the case per biblical history). These numbers differed by about a factor of 100,000. Subsequently, additional crystals were obtained and sent to an independent laboratory to have the rate of diffusion actually measured. The results agreed with the predictions based on an age of about 6,000 years. As with the ^{14}C in coal, numerous attempts have been made to explain away these data but none survive detailed examination.[47]

If radiometric dating is being relied on to irrefutably establish the "deep time" that evolution requires, then evolution does, indeed, have a problem, because careful examination of the evidence from radiometric dating indicates that the Earth is young.

Evolutionists still puzzle over speciation

A response by Dr Don Batten

> *"...the popular view of evolution is upside down. People think evolutionary changes are imperceptible in the short term but add up to big changes over millions of years. In fact, the opposite is true. It now appears that organisms evolve very rapidly in response to any changes in the environment, but in the longer term most evolutionary changes cancel each other out. So the longer the period you look at, the slower evolution appears" (p. 147).*

The "popular view of evolution" was Darwin's view, and a major prediction of this hypothesis. Furthermore, it has been the view for most of the time since Darwin. However, because the fossil record does not show this expectation (another 'Fail' for evolutionary notions), *New Scientist* re-writes history to try to shore up the idea. The fossils don't show a simple life form radiating out into all the varieties we see today, but they show the sudden appearance of all the major phyla (extreme diversity), in what evolutionists themselves call 'the Cambrian Explosion.' This is the opposite of evolutionary expectations and provides a strong refutation of the whole of evolutionary theory. This is a grand 'Fail' for evolution.

The book briefly mentions many examples of 'rapid evolution', without much explanation, but all of them are merely variations *within* a created kind of creature. The authors seem to be totally confused about adaptation within the limits of an existing creature's genome, versus the origin of novel features in new genomes so as to explain the origin of new body plans. Evolution has to explain the latter and none of the examples provided give evidence of 'evolution' doing that. For example, insecticide resistance is given as an example of rapid evolution. However, in no case has anyone found an example of insecticide resistance being due to a new gene being created by 'natural selection' (or mutations, for that matter, the only evolutionary means of getting new genes, but which don't seem to get much attention from these authors).[48]

The authors say, "in the longer term most evolutionary changes cancel each other out. So the longer the period you look at, the slower evolution appears." Indeed so. Using evolutionary notions of time, we see profound evidence of stasis (things staying much the same), not evolution . There are many examples of creatures unchanged over even hundreds of millions of years, according to evolutionary dating. Still-living examples include cyanophytes, jellyfish, starfish, brachiopods, sea urchins, etc.[49] Evolution is supposed to change one kind of creature into another, which is contradicted by the observation of so many examples of organisms remaining pretty much the same. The fossil record shows us organisms appearing and disappearing, but not changing from one kind into another, because evolution never happened.

Fig. 20. A fossil sea pen, Charniodiscus arboreus, *from the South Australian Museum, Adelaide (©sAVC Inc., photo by Dr Carl Werner) and a living sea pen,* Sarcoptilus grandis, *from Doubtful Sound, New Zealand, (©sOceanwideImages.com).*

> *"Although Darwin titled his book* On the Origin of Species, *speciation was one thing he could not explain. He called it the 'mystery of mysteries', and even a century and a half later the mechanism by which two groups of animals become genetically incompatible remains one of the greatest puzzles in biology" (p. 29).*

So, evolution is supposed to explain the diversification of life (all life forms)—the origin of species—from one or a few original life forms, but after a century and a half the origin of diverse new species remains "one of the greatest puzzles in biology." And yet, evolution is an unassailable fact in the minds of these authors?

This section of the book alludes to the peacock's tail, which Darwin used to support his ideas of sexual selection (12 years

Fig. 21. Close-up of a peacock's tail (photo by Magda Ehlers, pexels.com).

after publication of *Origin*). Peahens supposedly preferred males with ostentatious tails, and so the males with the most impressive tails passed on their 'impressive tail' genes to their offspring, thus impressive tails proliferated and got more impressive. The 'Just so' story supposedly explained the elaborate tail of the peacock. However, experiments by Japanese biologists have shown that mating success is unaffected by the ornateness of tails.[50] Darwinism fail!

Furthermore, the *design* of the peacock's tail is incredibly intricate, and a real problem for evolutionary story-telling.[51] Natural selection of existing ornate tails would not explain the origin of the genetic instructions needed to make such tails. A selection process can only select from what is already there; it cannot *create* something. This is just the tip of the iceberg in terms of what has to be explained in living things. Every basic type of organism has orphan genes, that is, genes that are not even similar to those in any possible evolutionary ancestors. Just one orphan gene is a fail for evolution because there is no possible mechanism for generating brand new genes that are not similar to existing genes.

Mechanisms of speciation are indeed very tricky to elucidate. However, while speciation was a total mystery to Darwin—and rapid speciation baffles evolutionists when observed today—it has long been an *expectation* of a biblical historical model of Earth history, as many organisms recolonized vacant ecological niches after the Flood.[52]

> *"And as charming as it sounds, there is no truth to the story that Darwin noticed the beaks of the finches [of the Galapagos Islands] were adapted to different diets and that this provoked his evolutionary theorizing. There was no Galapagos eureka moment" (p. 6).*

This is quite true, but whether the story is accurate regarding Darwin is irrelevant (a red herring). Textbooks the world over use this story to indoctrinate students with evolutionary notions of the origin of species, and therefore the story needs to be critiqued. And studies of the finches show that the evolutionary story does not add up:

1. The different 'species' of finches readily interbreed. Thus, the species are not even isolated reproductively, so strictly speaking they are not even species. There has been *no speciation*; i.e. true, reproductively isolated populations have not arisen! Mind you, creationist biologists have no problem with speciation, because the processes involve variation *within* created kinds, not the origination of new kinds.

2. The populations of the finches with differing beak shapes vary from season to season depending on the availability of different types of food, but the finches remain finches; adaptation and natural selection are operating here, which are not to be equated with evolution. They are not changing into something else. The limited variation observed within a created kind is consistent with biblical creation which logically presupposes a built-in ability to adapt, but does nothing to support grand-scheme evolution (e.g. that *T. rex*-like creatures changed into finches, and all the other bird types, over tens of millions of years).[53]

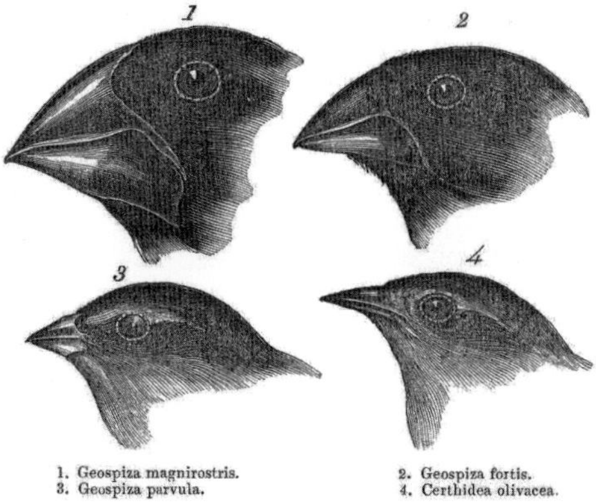

Fig. 22. Darwin's drawings of the contrasting beaks of four Galápagos finches, from his Journal of Researches *(2nd ed.), p. 379, 1845.*

Genes are not a simple blueprint for life

A response by Dr Peter Borger

> *"Some biologists are trying to shoehorn the new knowledge [advances in genetics, epigenetics and developmental biology] into traditional evolutionary thinking. Others ... believe a more radical approach may be required. We don't deny the roles of genetic inheritance and natural selection, but think we should look at evolution in a markedly different way. It is time for the theory of evolution to evolve" (p. 229).*

When I first read the title of this book, *How Evolution Explains Everything About Life*, I could not suppress a chuckle. In my own book *Darwin Revisited*, I spent a paragraph or two clarifying why theories that explain everything about anything are non-scientific. I quoted science philosopher Carl Popper: "the criterion of the scientific status of a theory is its falsifiability, or refutability, or testability [...]. A theory, which is not refutable by any conceivable event, is non-scientific.[54] Irrefutability is not a virtue of a theory (as people often think) but a vice."[55] Following Popper's criterion, if evolution explains *everything* about life, it cannot be a scientific theory, because *nothing* can contradict it.

Isn't it ironic that a panel of evolutionary scientists chose an anti-science title for their book? Still, the book includes an interesting set of statements—specifically those, like the one above (from the final chapter) addressing the future of evolution. Modern genetics has dramatically changed our understanding of what life is.

Until a few decades ago, it was possible to regard life as an utterly materialistic phenomenon. Indeed, the idea of evolutionary science has always been that, given enough time and energy,

Fig. 23. James Watson (middle) and Francis Crick (right) in 1962 when, along with Maurice Wilkins, they received the Nobel Prize for having elucidated the structure of DNA in 1953 (Marjorie McCarty, CC BY 2.5).

lifeless matter will turn into living, reproducing cells. Since 1953 we have known this cannot be the case. After Watson and Crick discovered the double helix structure of the DNA molecule, they embraced the idea "that the precise sequence of the bases [in the DNA] is the code which carries the genetical information".[56] The observation that biological systems are based on information, combined with the insight that information is an immaterial entity, should have led to a novel understanding of biology, as it has severe implications for origin theories. But it did not. The commitment to evolutionary materialism was far more important than scientific reality.

Our 21st century advances in genetic knowledge now really demand a rethink. And this is what the writers of this chapter of HEEEAL realize. There is far more information in the genome than just the protein code envisioned by Watson

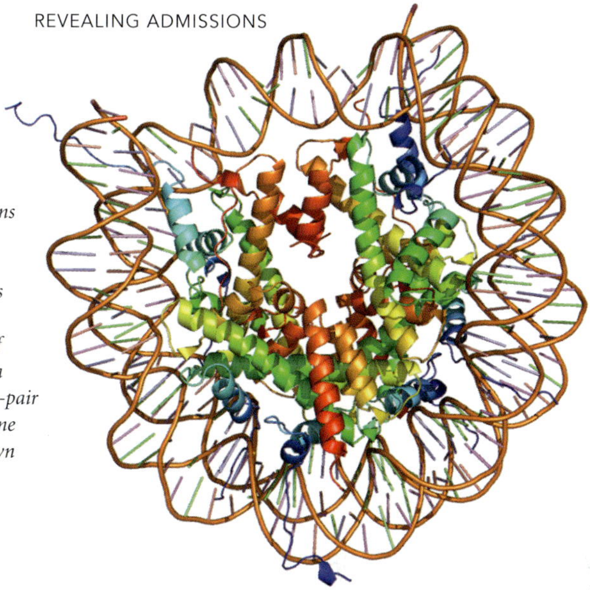

Fig. 24. Histones are proteins in the nuclei of eukaryotic cells around which DNA is packed into structural units called nucleosomes. The image is a 3-D rendering of the complex between such a nucleosome and a 146 base-pair DNA fragment, the backbone of which is depicted in brown (Emw, CC BY-SA 3.0).

and Crick in 1953. We now know of activator and coactivator codes for transcription and translation, mRNA splicing codes, and at least a dozen codes for novel regulatory RNAs. We find structural codes for genome stability, for spatial and temporal configuration of the chromosomes, for DNA recombination, and for inter-chromosomal communication. There is code for DNA modifications, for trafficking, for genetic switches, for non-random mutations, for gene editing, for integration and excision of transposons, for adaptations. On top of that we find the histone[57] code for epigenetic regulation of gene programs, as well as sugar codes and electrical codes for embryological development. Everything in living cells is controlled and regulated by coded information. The cell is a computer, which responds and adapts to the environment because it is programmed that way. Current genome biologists consider almost the entire genome as functional code, as the ENCODE Project Consortium demonstrated, 2012.[58]

Therefore, I fully agree with this statement in HEEEAL, but as long as the evolving evolution theory fails to develop to include an information concept, it cannot explain origins at all. Not the origin of genes. Not of cells. And not of organisms. Do the authors anticipate an information theory of biology? No. And therefore their musings remain futile.

> *"We now know that things other than genes are transmitted from parents to offspring. These include components of the egg, hormones, symbionts (microorganisms that live inside bodies), epigenetic marks (compounds that bind to DNA and turn genes on or off), antibodies, ecological resources and learned knowledge. ... It is time to let go of the idea that the genes we inherit are a blueprint to build our bodies" (pp. 230–231).*

The real irony is that they fail to see that the novel (epi)genetic mechanisms, which we now realize facilitate adaptations and speciation events, actually negate evolution; the latter were the reason why Darwin even invented evolution theory at all. The highly adaptable genomes of which our age is now aware reduce his grand idea of evolution to a non-sequitur, a fallacious remnant of the ignorant, pre-genetic era. That the book is published by the *New Scientist* does not really come as a surprise. *New Scientist* is a popular science magazine, which repeatedly presents opinions and hypotheses of evolutionists as if they were scientific realities. Usually there is more speculation than there are facts. The public may not be able to discriminate between opinions of scientists and science, but many scientists—like myself—can clearly tell the difference. HEEEAL? It would be more accurately titled, 'How Evolution Explains Everything *Apart from* Life'—in other words, not much at all.

Evolution still evolving—really?

A response by Dr Samuel Gan

> *"Organisms are not the 'throwaway survival machines' envisaged by Richard Dawkins and others, but instead often take the lead in their own evolution, dragging genetic change along in their wake. Move over selfish gene, and make way for the orchestrating organism" (p. 238).*

How many times has the Word of God needed revising? Not once (Matthew 24:35). Not once did the Bible record a 'slip of the tongue' or something that had to be re-written. Compared to the Bible, however, the authors of HEEEAL suggest that, for the umpteenth time, the evolution theory needs to be revised in the light of new information (e.g. molecular biology, epigenetics, etc.), maybe towards an 'Extended evolutionary synthesis'. While this might be conceded if the revision was certain to make evolution fool proof (it does nothing of the sort of course), the proposed incorporation of new concepts into the old way of thinking highlights their failure to acknowledge that evolution is fatally flawed. It seems that they're clutching at reasons to continue rejecting the biblical account.

Firstly, the authors claim that the organisms "take the lead in their own evolution". Here they are fallaciously presuming that organisms are able (somehow) to react sufficiently quickly to adapt to oncoming selection pressures, and even to orchestrate the required adaptations. Considering the parts required to build an eye, for example, what would the selection pressures be for

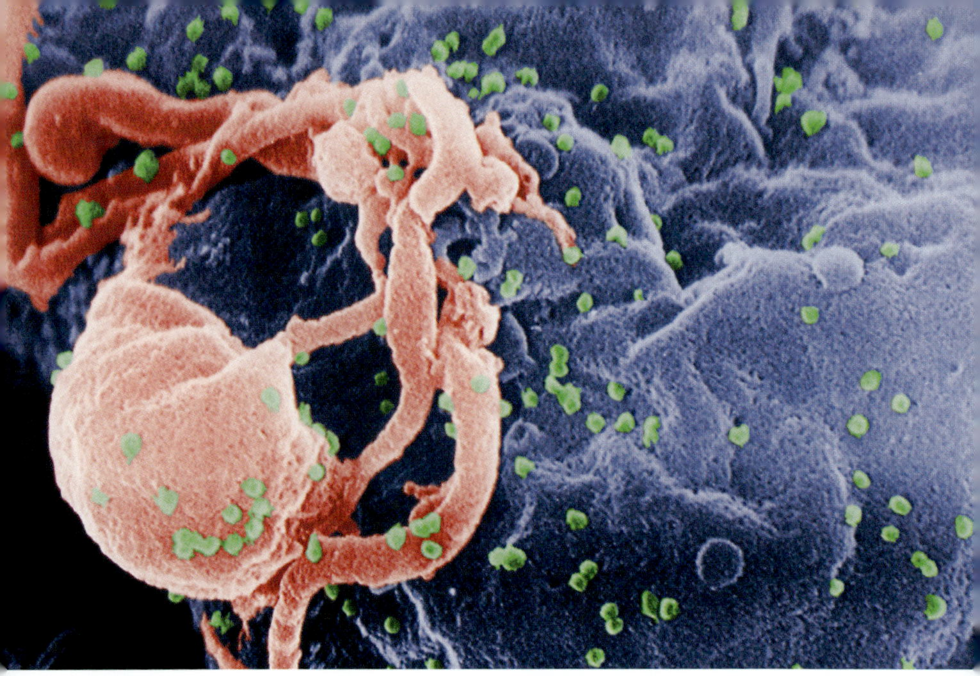

Fig. 25. False colour scanning electron micrograph of an infected lymphocyte (a blood cell; shown in blue) infected with HIV-1, shown in green (C. Goldsmith, Wikipedia).

every small step of the evolving 'proto-eye'? For organisms to evolve 'proactively' and slowly (without selection-pressures for intermediate steps), they would need to be sentient and self-conscious, and able to comprehend the required steps towards an eye, and foresee that having vision would be important or somehow advantageous. Explaining the evolution of vision itself is much less challenging compared to how such intelligence or self-awareness could evolve. The idea is preposterous. It is not even within the capacity of the conscious and sentient human to decide what we might wish to evolve (e.g. more hands, more eyes, or even simple things like better health or resistance to diseases), much less to dictate the passing of such desirable traits to our offspring (without extensive genetic engineering methods).

Secondly, there seems to be a failure on the authors' part to differentiate behavioural or physical changes from the genetic changes needed in evolution. Deciding to change one's dietary preference or engaging in some self-preserving behaviour does not create any new heritable genes that are required for evolutionary change. In other words, volition alone cannot alter our genetics, nor can we ensure that any modifications are passed on at procreation. Neither knowledge nor declarative memory[59] are inheritable or we would not need to learn a language from scratch with every generation.

Even supposing a proactive speciation adaptation could arise from the generation of variety, under extreme selection pressures, such 'pre-adaptation' would be severely limited, if even possible. Evidence for this can be seen using the HIV[60] model of rapid adaptation, which is fairly successful against intense selection pressures of the rapidly adapting host immune system, together with intense drug-related selection.[61] Despite years of intelligent human interventions with drugs, HIV has not evolved into a superior virus kind. In fact, the virus is further restricted by the overlapping genes within its genome, which make it next to impossible for random mutations to improve one gene without severely damaging the overlapping gene.

Thirdly, overlapping genes are also a problem in 'higher' organisms like mammals, with at least 10% of their genes overlapping, and many of them playing important roles such as tumour suppression and the determination of core features. It is for such reasons that, "Scientists who study overlapping genes use words like 'novel,' 'remarkable,' 'striking' and 'statistically improbable' to describe them."[62] Note the word "improbable" and the important point that the overlapping genes mean that mutations, whether random or intended, are very unlikely to leave crucial life processes undamaged, much less improve them.

Lastly, mutations are not as random as early, traditional Darwinists presumed them to be,[63] and this imposes a limit on the genetic variety an organism can generate to pre-emptively or intelligently prepare for oncoming selection pressures.[64] It is too simplistic to claim that the 'new synthesis' (neo-Darwinism; i.e. natural selection plus mutations) sufficiently explains the generational changes leading to evolutionary change (within the limited time-frame and frequent suddenness of the onset of selection pressures). This is the case *even if* epigenetics, evo-devo (or any of the other modern modifications of the standard evolutionary model) are thrown in to save evolution (in an 'extended evolutionary synthesis')—and it is regardless of whether the organism is passive or active (as HEEEAL's authors contend) in its own evolution.

> *"Innumerable examples of evolution in action can be seen all around us ... Evolution is as firmly established a scientific fact as the roundness of the Earth" (p. 245; repeated from this article's introductory remarks).*

The roundness of the Earth is affirmed in the Bible (Isaiah 40:22), as is the 6-day creation account. Since the writing of the Pentateuch (including Genesis) in 1,486 BC or thereabouts, there has been no 'proof' against the Bible that could not be thoroughly counteracted (and that is certainly not from a lack of skeptics!). The roundness (actually sphericity) of the Earth has been a verifiable, testable phenomenon, through countless lines of enquiry. But who has observed a change in biological kind (such as an ape into a human), much less in a scientifically reproducible manner or in such a way as to uncover the mechanisms? Even in experiments involving intense selection pressures (e.g. directed evolution experiments), the drastic genetic changes required for a change in kind are still not observed.

Without such scientific evidence, it is vain to expect evolution (in whatever modernized, updated version) to explain the world around us—in fact, such a view is itself a position of faith. It is not that the act of faith itself should be shunned, but rather its object. The misplaced faith in big-picture evolution is faith in man's own hypothesis, one that is replete with unreasonable biases and assumptions. This idea is as far "as the east is from the west"[65] from a biblically-based faith in an all-powerful, intelligent Creator—the One Who reveals His creative work in the very Scriptures that many have tried to cast doubt upon since the beginning of Creation.

What do you, the reader, put your faith in? A continuously revised ideology stemming from assumptions, fallacies, and biases? Or the irrefutable, enduring account of the all-powerful Creator that was given to man millennia ago, and is still standing the test of time? Compared to the opinions of the writers of HEEEAL, the Bible is the testimony of the Creator Who made everything that exists, both seen and unseen. It also speaks of the extent to which He went to rescue creatures like us, of dying a wretched death by man's hands, crucified (then raised again) to redeem all of creation.

Conclusions

By Philip Bell

We noted at the very start that HEEEAL carries the strikingly up-beat subtitle: *From Darwin's brilliant idea to today's epic theory.* Is this justified, in the light of what you have now read? We hope that, having perused the carefully reasoned responses by our panel of 10 Ph.D. scientists, you will concur that such optimism is misplaced at best. Perhaps you would agree that the subtitle is actually delusional, flying in the face of stark scientific realities.

While this booklet ostensibly pitted two groups of Ph.D. scientists against one another (from different sides of the ideological debate), please do not make the mistake of seeing it as merely a sort of 'playground scrap' between big bruisers. There is a danger

Fig. 26. Which side has the strongest, most rational arguments? (pexels.com)

that, in seeing such heavy-hitters going head-to-head , people on the side-lines can fall back on the argument from authority. In other words, we can find ourselves asking, 'Which scientists are the most well-known? Who have the most impressive credentials? …the most published papers? …the most books to their names?' and so on. However, that is no way to judge points of scientific contention, let alone a question that purportedly relates to "everything about life".

Does evolution explain everything about life? Really? Our team of scientists have argued squarely within their specialist fields, concluding that the answer is a resounding 'No'. They have made this case based on actual scientific evidence, rather than offering their opinions or assertions. However, they would definitely not expect you, the reader, to unquestioningly accept what they've said on the basis of their scientific credentials—that would be a misplaced authority—rather, on the strength of their rational, scientific arguments.

Unfortunately, it is our experience that the generally expected standards of scientific integrity are often flouted in popular level presentations of evolution—so much so that what is served up is essentially propaganda, with a good dose of 'trust us, we're the experts' thrown in. Ironically, the very title of the *New Scientist* book critiqued here is itself completely unscientific—it smacks of scientism:

> Scientism is not the same thing as science. Science is a blessing, but scientism is a curse. Science, I mean what practicing scientists actually do, is acutely and admirably aware of its limits, and humbly admits to the provisional character of its conclusions; but scientism is dogmatic, and peddles certainties. It is always at the ready with the solution to every problem, because *it believes that the solution to every*

> *problem is a scientific one*, and so it gives scientific answers to non-scientific questions. Owing to *its preference for totalistic explanation*, scientism transforms science into an ideology, which is of course a betrayal of the experimental and empirical spirit [emphases added].[66]

Scientific knowledge should be established by experiment or demonstration. All arguments should be evidence-based. Arguments from authority and the sort of philosophical overreach inherent within *How Evolution Explains Everything About Life* are the antithesis of true science. If you compare the book's title with the italicised comments in the previous quotation (written by a confessed secular humanist), the book's undisguised ideological nature is incontestable.

Is it not true that the respondents in this booklet also have an ideological commitment? Absolutely, and that was made clear at the outset—they are all subscribe to a biblical creation viewpoint. However, *all of us* subscribe to a worldview (even if subconsciously), laypeople and scientists included. But which ideology (set of beliefs) is under threat from the hard facts of science and which is reinforced by those same facts? Those who take a long hard look at the many lines of scientific evidence are able to make an informed decision. But even that decision is not made in a spiritual vacuum.

The question asked throughout this booklet has been, whether HEEEAL lives up to its grand claims. Or to use the metaphor which I posed at the start, the real proof of the pudding is in the eating—and now that you've done so, just how grand a pudding is it? That is, now that you've read each Ph.D. scientist's cogent engagement with various inflated claims and telling admissions, is the title of HEEEAL justified? Or is it not the case that their acknowledgement that evolutionary theory needs a substantial makeover belies the very title of the book? You must decide.

When all is said and done, the age-old debate with which we have been engaged here really boils down to an issue of trust. Who do you trust? When there is a clear conflict between the *interpretations* of the scientific data—big-picture evolution-believers versus skeptics (in this case biblical creationists)—whose word will you accept? If that sounds like a contradiction of what was previously said about authority, then ask yourself instead, 'Am I really prepared to go where the evidence leads me, *regardless* of the opinions of other human beings?' The advice of wise King Solomon (c. 970–931 BC) has stood the test of time:

"Trust in the Lord with all your heart, and do not lean on your own understanding. In all your ways acknowledge him, and he will make straight your paths."

Proverbs 3:5–6

Fig. 27. Evolution answers none of the really important questions about life (photo by Lukas Rodriguez, pexels.com, CC0).

Finally, while evolution does *not* "explain everything about life", there *are* answers to life's biggest and most pressing questions—not only 'Where do I come from?' (the question of origins) but more importantly such questions as these :

- Why am I here? What is the meaning of my existence?
- Why the aching sense that there is more to life than this material world?
- What is true happiness? Can this really be attained and, if so, how?
- Am I loved? Is there an escape from hopelessness?
- Why is there so much suffering, misery, and death in the world?
- What happens after death? How can I escape the fear of death?

I know I speak for each of the scientists who have contributed to this response in saying that our chief desire is for readers of this booklet to find answers to these questions 'about life'. We believe wholeheartedly that the answers are to be found through a personal relationship with the Creator of us all—and through Him alone. No self-help philosophy or adherence to man-made religion will be of any help to you. The Bible is a true, authentic, historically accurate, reliable, and trustworthy revelation from God. And the way to true fulfilment, lasting peace, and joy in life is *a Person*. Jesus said:

> *"I am the way, and the truth, and the life. No one comes to the Father except through me."* John 14:6

> *"Come to me, all who labour and are heavy laden, and I will give you rest."* Matthew 11:28

Fig. 28. The Word of God reveals who Jesus is and explains everything you need to know about life.

Coming to Jesus Christ means a person chooses repentance. This means turning from selfishness and from recklessly living without resort to God. It also means thanking Him in faith that, just as the Bible teaches, His sacrifice on the cross has paid your debt of sin and guilt in full (a debt that cannot be cancelled in any other way):

> *"The times of ignorance God overlooked, but now he commands all people everywhere to repent, because he has fixed a day on which he will judge the world in righteousness by a man whom he has appointed; and of this he has given assurance to all by raising him from the dead."* Acts 17:30–31

It really is true that the Bible explains everything important that you need to know about life, and to know Jesus is to actually *know* "the life".

References and notes

1 *How Evolution Explains Everything About Life: From Darwin's brilliant idea to today's epic theory*, 2017; published in Great Britain by John Murray Learning, London; in the United States by Nicholas Brealey Publishing, Boston, MA.

2 Fifteen more "writers and editors" were also involved, ref. 1, p. 8.

3 Carter, R. (Ed.), *Evolution's Achilles' Heels*, Creation Book Publishers, Powder Springs, GA, USA, 2014.

4 Molecules in living organisms, such as sugars and amino acids, can be right- or left-handed. Mirror images of one another (stereoisomers), such molecules are not superimposable (like left and right gloves). Homochiral means the same handedness.

5 It may also be spelled hydroxyl.

6 Allentoft, M.E. *et al.*, The half-life of DNA in bone: measuring decay kinetics in 158 dated fossils, *Proc. Royal Society B* **279**(1748):4724–4733, 7 December 2012 | doi:10.1098/ rspb.2012.1745.

7 Bernhardt, H.S., The RNA world hypothesis: the worst theory of the early evolution of life (except for all the others), *Biology Direct* **7**:23, 13 July 2012.

8 According to the National Human Genome Research Institute (part of the US's *National Institutes of Health*), "there are probably about 20,500 human genes"; **genome.gov/human-genome-project/What**, accessed 6 May 2020.

9 Mattick, J., cited in: Gibbs, W.W., The unseen genome: gems among the junk, *Scientific American* **289**(5):26–33, November 2003; pp. 29–30.

10 Batten, D., Dazzling DNA: Huge study highlights stupendous design in human DNA, *Creation* **35**(1):38, January 2013; **creation.com/dazzling-dna**.

11 Gates, B., *The Road Ahead*, Penguin Group, New York, p. 188, 1995.

12 Carter, R., The four dimensional human genome defies naturalistic explanations, **creation.com/four-dimensional-genome**, 6 October 2016.

13 Buschauer R. *et al.*, The Ccr4-Not complex monitors the translating

ribosome for codon optimality. *Science* **368**(6488):281, 2020.

14 Carter, R., More evidence for the reality of genetic entropy, *Journal of Creation* **28**(1):16–17, 2014; **creation.com/evidence-for-genetic-entropy**.

15 Marin, M. *et al.* Protein folding and tRNA biology, *Biophys Rev* **9**(5):573–588, 2017.

16 Homology and Embryology Questions and Answers (FAQ page), **creation.com/embryonic**.

17 'Vestigial' Organs Questions and Answers (FAQ page), **creation.com/vestigialorgans**.

18 Carter, R., The slow, painful death of junk DNA, *Journal of Creation* **23**(3):12–13, 2009; **creation.com/junk-dna-slow-death**.

19 In computing terms, ROM = read only memory (storage), RAM = random access memory (volatile memory).

20 Liu, Y. and Moran, D., Do new functions arise by gene duplication? *Journal of Creation* **20**(2):82–89, 2006; **creation.com/gene-duplication2**.

21 Carter, R., Can biologically active sequences come from random DNA? *Journal of Creation* **31**(3):82–89, 2017; **creation.com/random-dna**.

22 Sanford, J. *et al.*, The waiting time problem in a model hominin population, *Theoretical Biology and Medical Modelling* **12**:18, 2015.

23 Lawton, G., Why Darwin was wrong about the tree of life, *New Scientist* **201**(2692):34–39, 24 January 2009.

24 Tomkins, J., and Bergman, J., Incomplete lineage sorting and other 'rogue' data fell the tree of life, *Journal of Creation* **27**(3):84-92, 2013; **creation.com/rogue-data**.

25 Sheridan, K., Huge viruses may open 'Pandora's box': French study, phys.org/news/2013-07-huge-viruses-pandora-french.html, 18 July 2013.

26 Brown University, Study rearranges some branches on animal tree of life, phys.org/news/2008-03-rearranges-animal-tree-life.html, 5 March 2008.

27 See: Jerlstrom, P., Is the evolutionary tree turning into a creationist orchard? *Journal of Creation* **14**(2):11–13, 2000; **creation.com/creationist-orchard**.

28 Vogel, S., *Cats' Paws and Catapults: Mechanical Worlds of Nature and People*, Penguin Books Ltd, London, p. 23, 1998.

29 Shu, D., *et al.*, Lower Cambrian vertebrates from south China, *Nature* **402**(6757):42–46, 1999.

30 Janvier, P., Catching the first fish, *Nature* **402**(6757):21–22, 1999.

31 Caldwell, M.W., Nydam, R.L., Palci, A., and Apesteguia, S., The oldest known snakes from the Middle Jurassic–Lower Cretaceous provide insights on snake evolution, *Nature communications* **6**(5996), 27 January 2015 doi | 10.1038/ncomms 6996.

32 Brussatte, S.L., Niedźwiedski, G., and Butler, R.J., Footprints pull origin and diversification of dinosaur stem lineage deep into Early Triassic, *Proceedings of the Royal Society B* **278**(1708):1107–1113, 2011.

33 Hochuli, P.A. and Feist-Burkhardt, S., Angiospermlike pollen and Afropollis from the Middle Triassic (Anisian) of the Germanic Basin (Northern Switzerland), *Frontiers in Plant Science* **4**(344): 1–14, 2013.

34 CNRS, Discovery of a complex, multicellular life from over two billion years ago, physorg.com, 30 June 2010.

35 Gingerich, P.D., Haq, M., Zalmout, I.S., Khan, I.H. and Malkani, M.S., Origin of whales from early artiodactyls: Hands and feet of Eocene Protocetidae from Pakistan, *Science* **293**(5538):2239–2242, 2001.

36 Walker, T., Learning the lessons of Mount St Helens, *Creation* **39**(3):23–27, July 2017; **creation.com/lessons-from-mount-st-helens**.

37 Walker, T., Polystrate fossils: evidence for a young earth, *Creation* **29**(3):54–55, June 2007; **creation.com/polystrate**.

38 Walker, T., Granite formation: catastrophic in its suddenness, *Journal of Creation* **21**(2):13–15, August 2007; **creation.com/granite**.

39 Batten, D., Age of the earth: 101 evidences for a young age of the earth and the universe, **creation.com/age**, updated 18 September 2019.

40 What about carbon dating? Chapter 4 of *The Creation Answers Book*, Creation Book Publishers, Powder Springs, GA, USA, 2014; **creation.com/images/pdfs/cabook/chapter4.pdf**.

41 Swenson, K., Radio-dating in rubble: The lava dome at Mount St Helens debunks dating methods, *Creation* **23**(3):23–25, June 2001; **creation.com/radio-dating-in-rubble**.

42 Snelling, A., Radioactive 'dating' failure: Recent New Zealand lava flows yield 'ages' of millions of years, *Creation* **22**(1):18–21, December 1999; **creation.com/radioactive-dating-failure**.

43 Snelling, A.A., Radioisotope dating of rocks in the Grand Canyon, *Creation* **27**(3):44–49, June 2005; **creation.com/dating-gc**.

44 Baumgardner, J., Carbon-14 evidence for a recent global flood and a young earth, chapter 14 of: Vardiman, L. *et al.* (Eds.), *RATE II: Radioisotopes and the Age of The Earth: Results of a Young-Earth Creationist Research Initiative (Volume II)*, Institute for Creation Research and the Creation Research Society, pp. 587–630, 2005; **icr.org/article/carbon-14-evidence-for-recent-global**.

45 Wieland, C., Radiocarbon in dino bones: International conference result censored, **creation.com/c14-dinos**, 22 January 2013.

46 Humphreys, D.R., Young helium diffusion age of zircons supports accelerated nuclear decay, in: Vardiman, L. *et al.* (Eds.), *Radioisotopes and the Age of The Earth: Results of a Young-Earth Creationist Research Initiative*, Volume II, Institute for Creation Research and the Creation Research Society, San Diego, CA, 2005.

47 Humphreys, R., Helium evidence for a young world continues to confound critics, **creation.com/helium-critics**, 29 November 2008.

48 For example, Catchpoole, D., Pesticide resistance is not evidence of evolution; **creation.com/pesticide**, 20 August 2009.

49 See articles on living fossils: **creation.com/topics/living-fossils.**

50 Takahashi, M. and others, Peahens do not prefer peacocks with more elaborate trains, *Animal Behaviour* **75**(4):1209–1219, 2008 | doi:10.1016/j.anbehav.2007.10.004.

51 Burgess, S., The beauty of the peacock tail and the problems with the theory of sexual selection, *Journal of Creation* **15**(2):94–102, 2001; **creation.com/peacock**.

52 Catchpoole, D. and Wieland, C., Speedy species surprise, *Creation* **23**(2):13–15, March 2001; **creation.com/speedy**.

53 Lightner, J.K., Unmasking natural selection—*A review of 40 Years of Evolution: Darwin's finches on Daphne Major Island* by Peter and Rosemary Grant, *Journal of Creation* **32**(2):37–41, 2018; **creation.com/unmasking-natural-selection.**

54 Borger, P., *Darwin Revisited, or how to understand biology in the 21st century*, Scholars' Press, 2018.

55 Popper, K.R., *Conjectures and Refutations: The Growth of Scientific Knowledge*, Routledge, London, p. 36, 1963.

56 Watson, J.D. and Crick, F.H.C., Molecular structure of Nucleic Acids: A structure for Deoxyribose Nucleic Acid, *Nature* **171**:737–738, 1953.

57 Histones are proteins that hold the DNA in chromosome structures.

58 ENCODE Project Consortium, An integrated encyclopedia of DNA elements in the human genome, *Nature* **489**:57–74, 2012.

59 Declarative memory is the conscious recollection of various facts or past events, thus is invariably long-term memory.

60 HIV is 'human immunodeficiency virus'; serious infection by HIV leads to AIDS (acquired immunodeficiency syndrome).

61 Yeo, J.Y., Goh, G-R., Su, C.T-T. and Gan, S.K-E., The determination of HIV-1 RT mutation rate, its possible allosteric effects, and its implications on drug resistance, *Viruses* **12**(3):297, 9 March 2020 | doi.org/10.3390/v12030297.

62 Armbrecht, H.J., Thank God for overlapping gene, Open Letters, *ASBMBTODAY*, asbmb.org, 1 April 2014.

63 Chan, K-F., Koukouravas, S., Yeo, J.Y., Koh, D.W-S. and Gan, S.K-E., Probability of change in life: Amino acid changes in single nucleotide substitutions, *Biosystems* **193–194**, #104135, June 2020 | doi.org/10.1016/j.biosystems.2020.104135.

64 Yeo, J.Y., Koh, D.W-S., Yap, P., Goh, G-R. and Gan, S.K-E., HIV-1 Spontaneous mutations in HIV-1 Gag, protease, RT p66 in the first replication cycle and how they appear: Insights from an *in vitro* BSL2 assay on mutation rates and types, bioRxiv – Microbiology, 27 July 2020 | doi.org/10.1101/679852.

65 To borrow from Psalm 103:12; it conveys an infinite distance between the two things under consideration.

66 Wieseltier, L., Perhaps culture is now the counterculture: A defense of the humanities, *The New Republic*, 19 May 2013; newrepublic.com/article/113299/leon-wieseltier-commencement-speech-brandeis-university-2013, accessed 22 July 2020.

RECOMMENDED RESOURCES

Evolution's Achilles' Heels

Edited by Dr Robert Carter, featuring nine Ph.D. scientists

Directly demolishes the very pillars of the belief system that underpins our now-secular culture—evolutionary naturalism. Nine scientists expose the fatal flaws of evolution on topics such as natural selection, the origin of life, geology, genetics, radiometric dating, the fossil record, cosmology, and ethics. In a nutshell, it's a comprehensive outreach tool like no other.

The Greatest Hoax on Earth?

Refuting Dawkins on evolution

Dr Jonathan Sarfati

Richard Dawkins, the undisputed high priest of evolution and atheism, says his book *The Greatest Show on Earth* is the first time he has presented all the evidence for evolution/long ages. Scientist and logician Dr Jonathan Sarfati, relentlessly demolishes Dawkin's claims point-by-point, showing biblical creation makes more sense of the evidence.

Evolution and the Christian Faith

Theistic evolution in the light of Scripture

Philip Bell

This book assesses theistic evolution against the backdrop of the whole gamut of Christian doctrines: the attributes of God, the deity of Jesus Christ, the problem of evil, the nature of mankind, original sin, death, the Gospel, the restored new creation, missions and evangelism.

A Question of Origins

Created or Evolved?

Shelby MacFarlane

This high school age textbook gives students the tools they need to evaluate where we, life, and the universe came from. Richly illustrated in full colour, it fairly presents what evolutionists believe and why, but then makes the powerful scientific case for creation—a case that is ignored by secular textbooks.

The Deep Time Deception

Examining the case for millions of years

Michael Oard

Is there room in the Bible for deep time? Where do we get the idea that the universe is billions of years old? Does 'science' prove the earth is old? What about radiometric dating? What about fossils? These questions, and many more, are tackled and clearly answered.

Defending Genesis

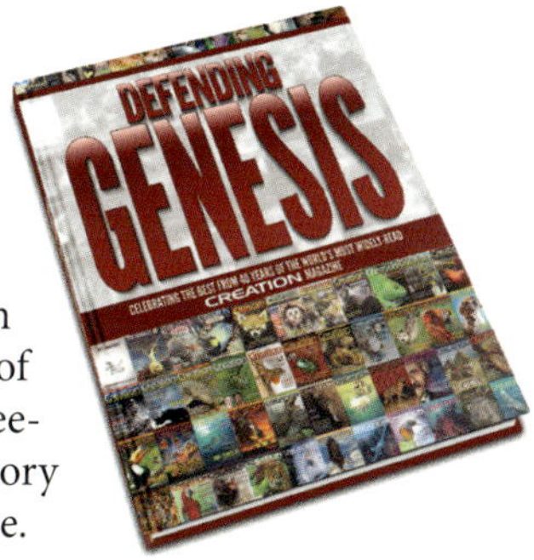

In 1978, there were few resources for Christians to defend biblical creation. That is why Carl Wieland founded *Creation* magazine, which would eventually become the most widely-read creation publication in the world. To celebrate 40 years of cutting-edge creation content, this beautiful coffee-table-style book collects articles from across the history of this ground-breaking publication into one volume.

For more information visit

CREATION.com

CREATION MINISTRIES INTERNATIONAL

AUSTRALIA
Creation Ministries International
P.O. Box 4545,
Eight Mile Plains, QLD 4113, Australia
Phone: (07) 3340 9888
Fax: (07) 3340 9889

CANADA
Creation Ministries International
300 Mill St, Unit 7,
Kitchener, ON, N2M 5G8
Phone: (519) 746-7616
Fax: (519) 746-7617
Subscriptions and orders only:
1-888-251-5360

NEW ZEALAND
Creation Ministries International
P.O. Box 13227, Onehunga,
Auckland 1643, New Zealand
Phone: (09) 537 4818
Fax: (09) 537 4813

SINGAPORE
Creation Ministries International
P.O. Box 195, 911207, Singapore
singapore@creation.info

SOUTH AFRICA
Creation Ministries International
P.O. Box 3349,
Durbanville 7551,
South Africa
Phone: (021) 979 0107
Fax: (086) 519 0555

UK and EUROPE
Creation Ministries International
15 Station Street,
Whetstone
Leicestershire, LE8 6JS, UK
Phone: 0116 2848 999

USA
Creation Ministries International
P.O. Box 350
Powder Springs,
GA 30127, USA
Phone: 800-616-1264
Fax: 770-439-9784

OTHER COUNTRIES
Please contact the Australian office.
aus@creation.info

For more information on creation/evolution and Bible-science issues

CREATION.com